DET 3470: CATIA V5 Tutorials

Table of content

Chapter 1 – Introduction to Catia V5..2

Chapter 2 – Sketcher Workbench..10

Chapter 3 – Part Design Workbench..24

Chapter 4 – Drafting Workbench...38

Chapter 5 – Assembly Design Workbench...48

Chapter 6 – Generative Shape Design Workbench..64

Chapter 7 – Rendering Workbench..91

Chapter 8 – Parametric Designing..97

Chapter 9 – Advanced Techniques...108

Chapter 1 – Introduction to Catia V5

CATIA is a powerful tool that is used in a wide range of industries, primarily aerospace and automotive. The CATIA acronym stands for: "Computer Aided Three Dimensional Interactive Application" and V5 stands for Version 5, and R20 stand for Release 20.

CATIA V5 Application Tools

In CATIA V5, each Application tool has several Workbenches. You need to select an appropriate Application tool to start modeling.

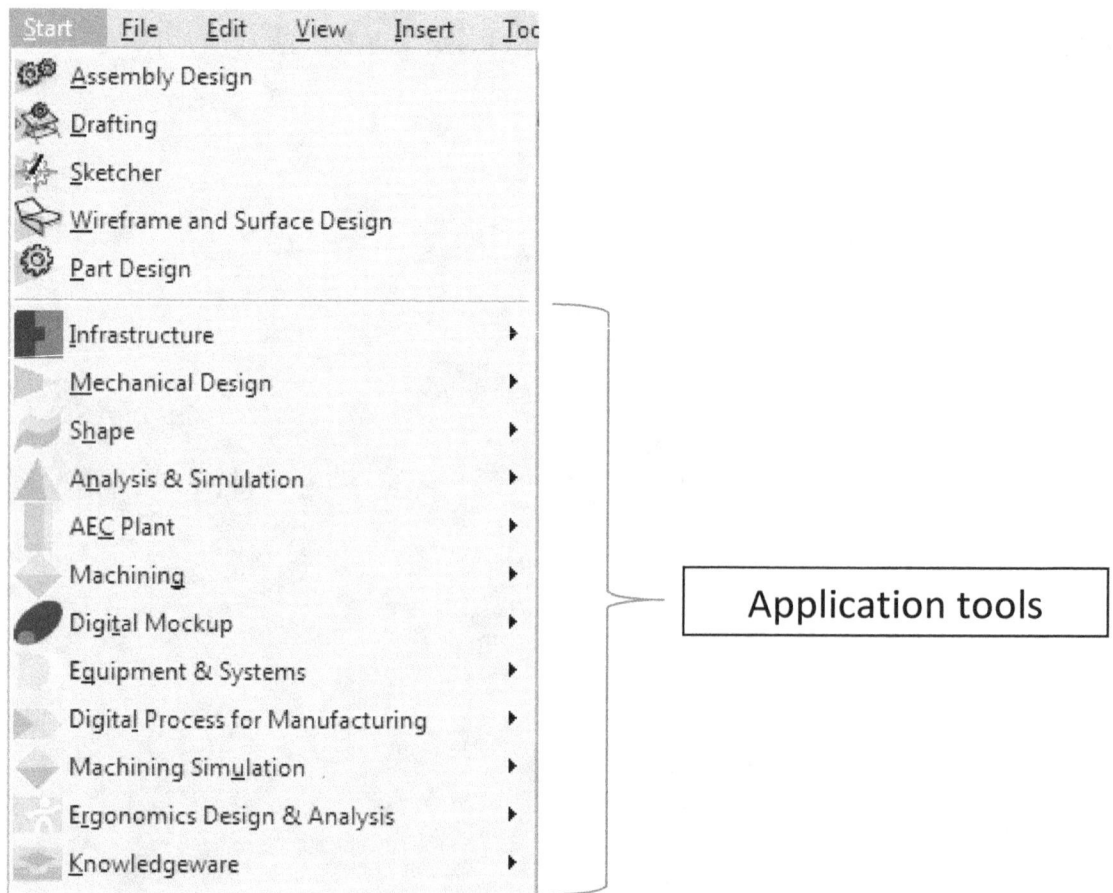

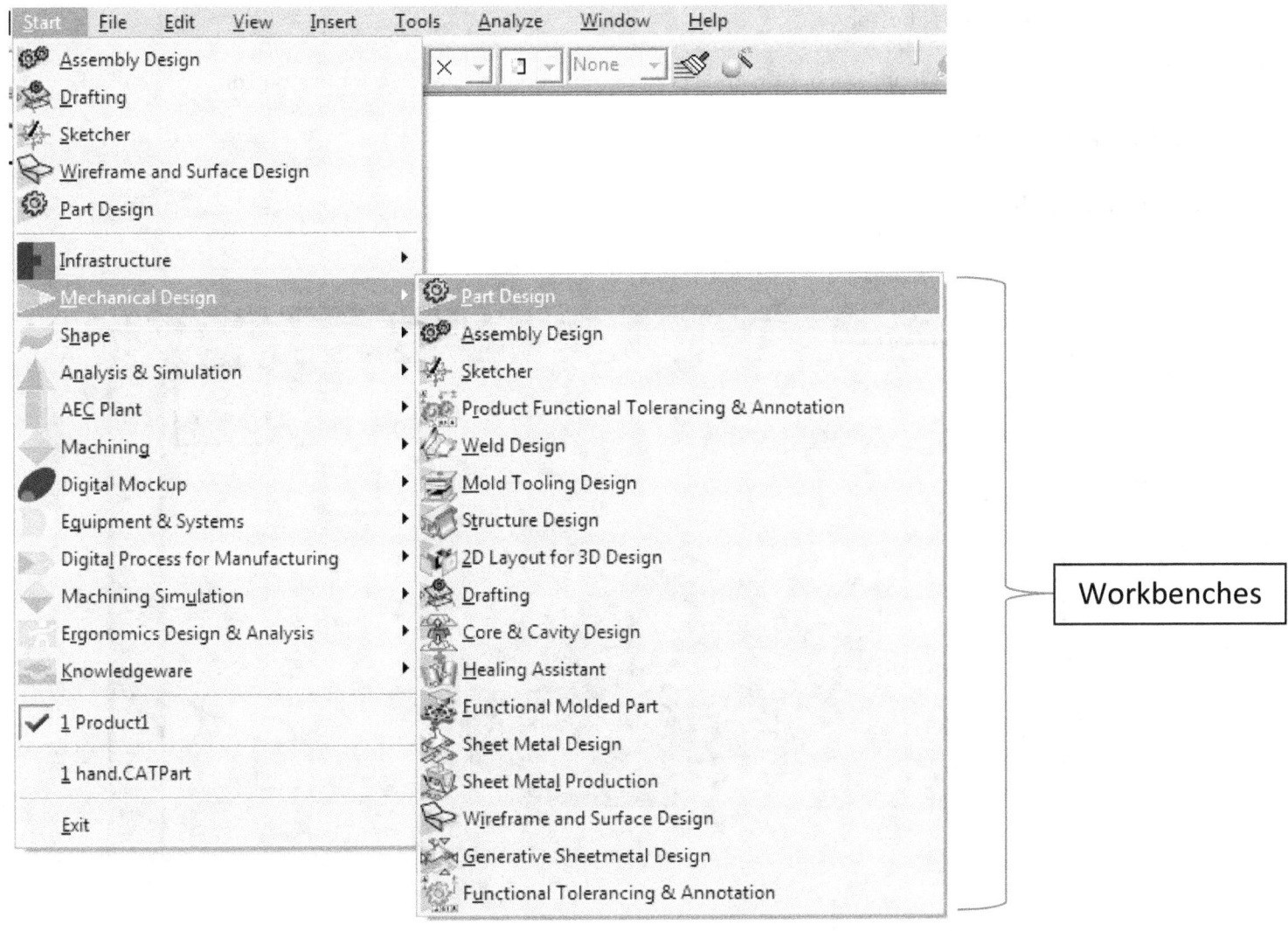

For DGET 3470; we are going to use the following Application tool and Workbenches:

1. Infrastructure Application tool
   - Photo Studio

2. Mechanical Design Application tool
   - Part Design
   - Assembly
   - Sketcher
   - Drafting

3. Shape Application Tool
   - Generative Shape Design

Manipulating mouse buttons in Catia V5

In CATIA V5, there is no need to use the keyboard to zoom in/out, Pan, or Rotate.

Pan: Hold down the middle mouse button

Zoom in/out: Hold down the middle mouse button and click one time on the right side mouse button

Rotate: Hold down both the middle and right side mouse buttons.

Standard Screen Layout

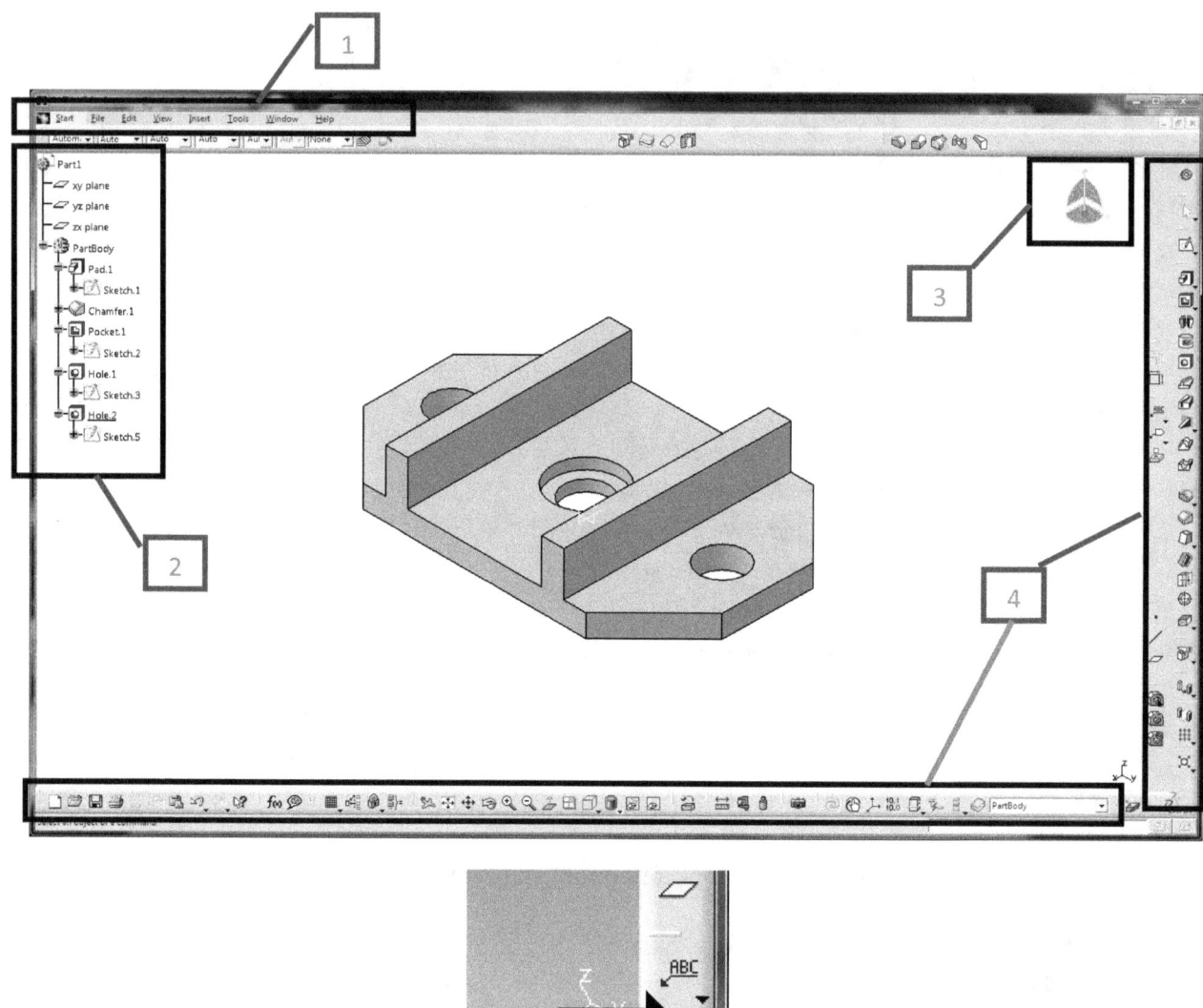

If there are more toolbars hiding, you can see double-arrows on the bottom as shown on above. Pull them out to see all the available toolbars.

1. Pull down menu: each menu has a wide range of functions and options as shown below.

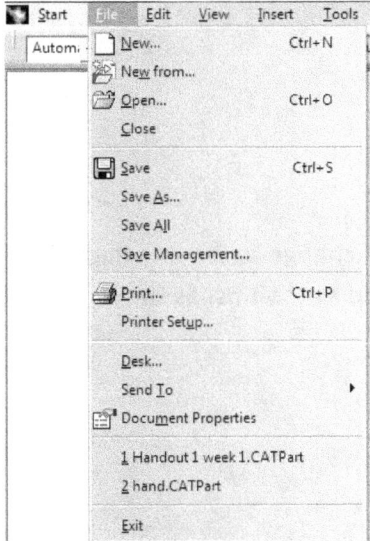

2. The specification tree: as the design progresses, this tree will contain the history of tools and processes used to create a part.

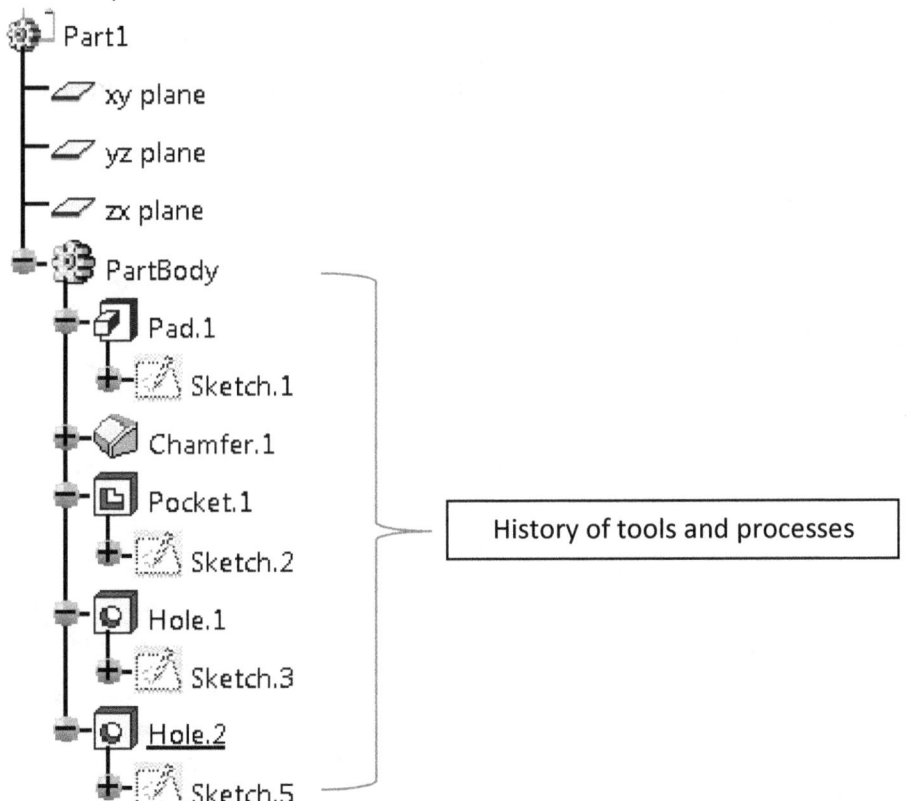

3. The compass: this can be used to move or rotate the part instead of mouse button manipulation. The compass can be used often in an assembly. When the compass is highlighted in green, it means a selected part can be

moved.

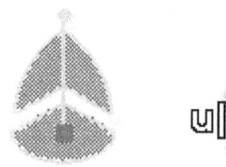

4. Toolbars: There are many toolbars located all around the screen. These will change based on which Workbench the user is in. Place the cursor over an icon, and it's function will be revealed as tooltips, as shown below.

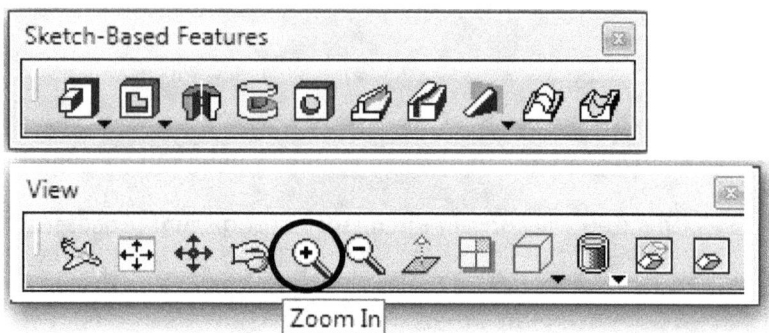

By clicking the empty area of toolbar, the list of activated toolbars appear. Necessary toolbars can be hidden or shown.

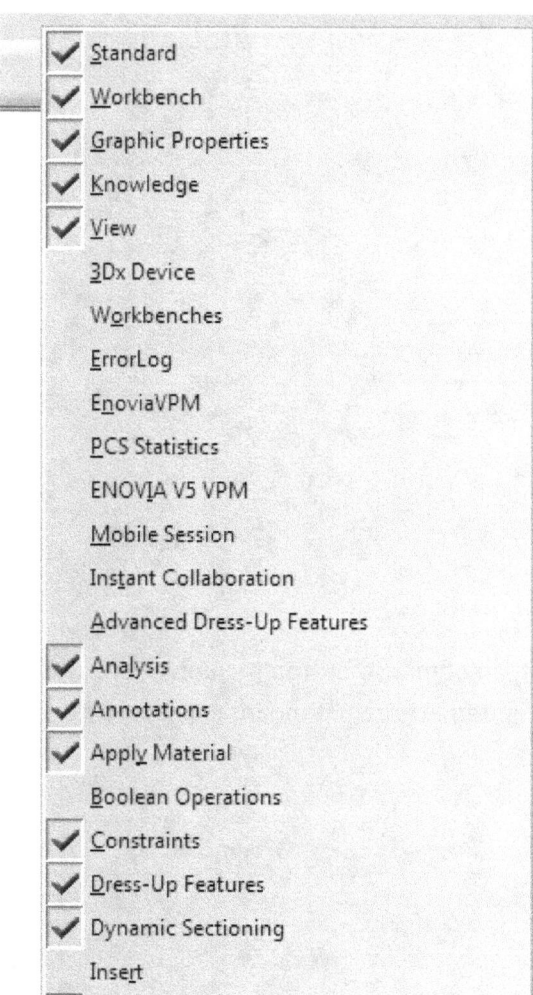

6

Click Customize to reveal this window. Under the Toolbars tab, toolbars can be controlled as preferred. For example, if one would like all the toolbars to be placed and locked, go to the Option tab and activate, "Lock Toolbar Position." This will keep toolbars from disappearing or moving.

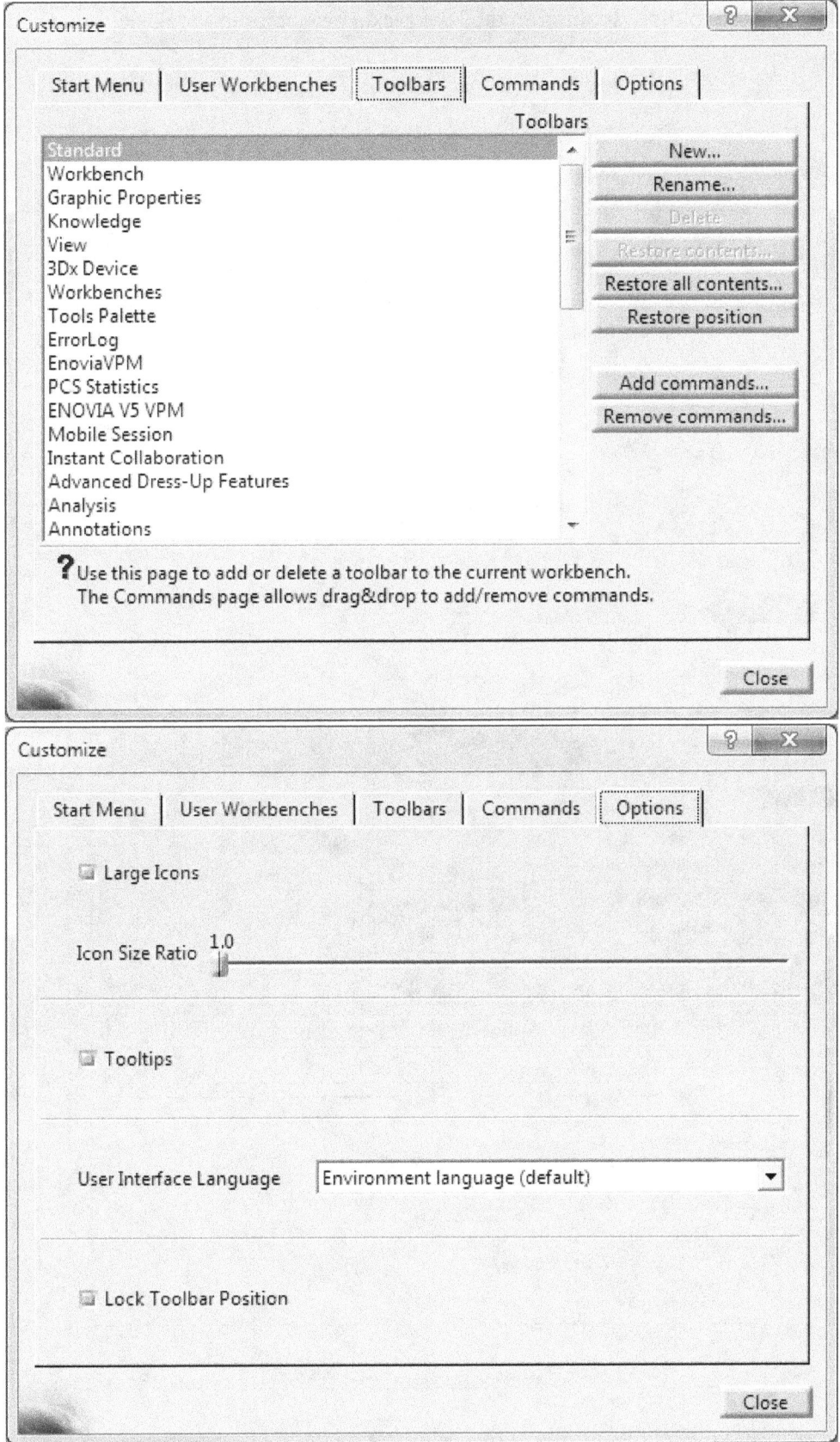

Editing set-up

In order to edit/change the default set-up, use Options under Tools as shown below. In the Options window, there are many options, and each option has several tabs. Select the correct option and tab to change/edit. To change the background, click Display on the left side, go to the Visualization tab, and pick a new color in the Background slot.

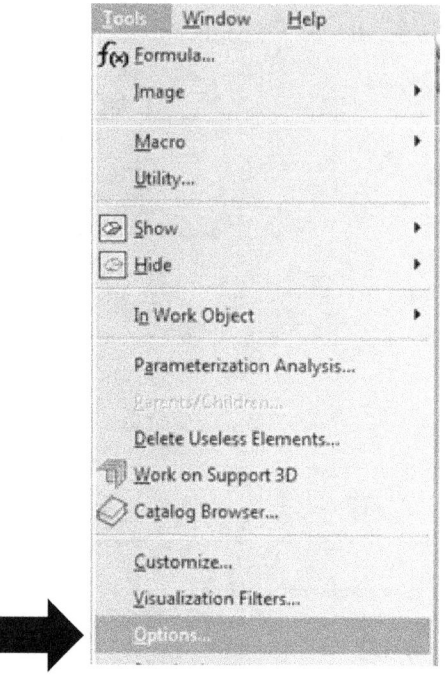

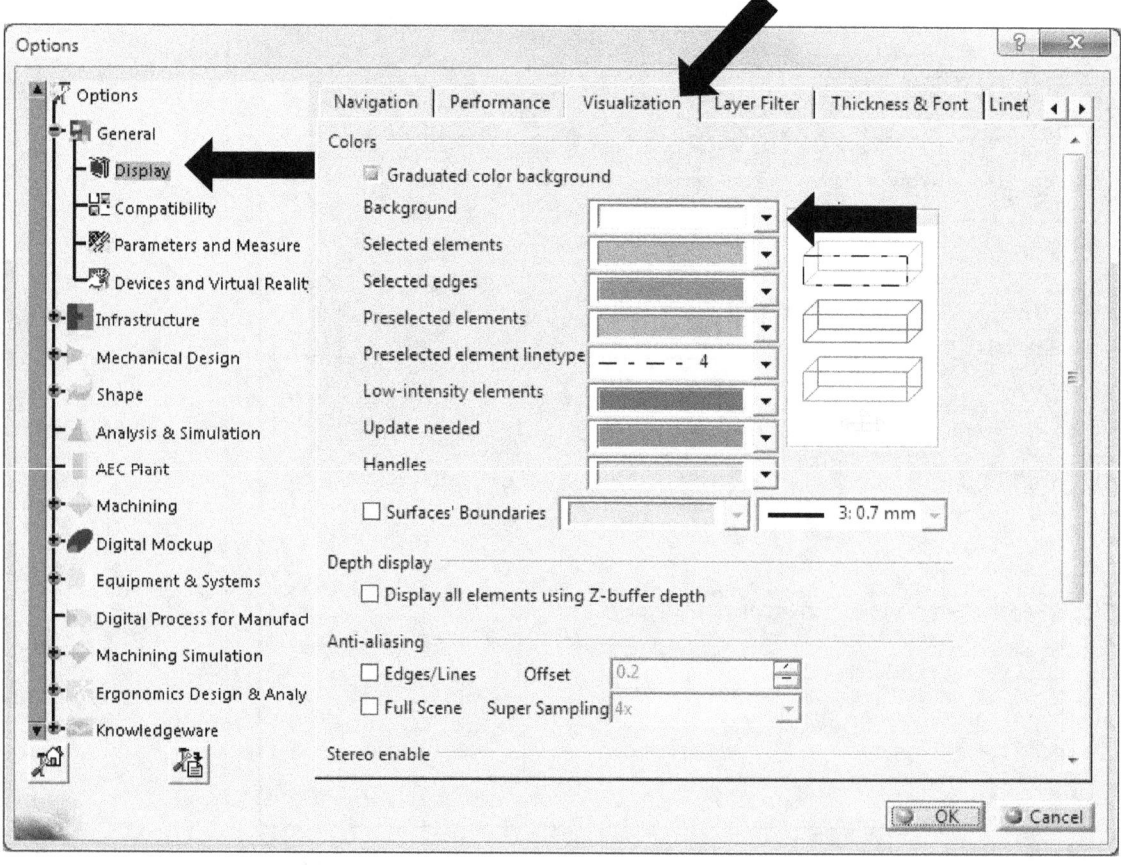

To change the unit (mm or inches) go to the Parameters and Measure on the left side, and in the Units tab, select Length. Choose the necessary unit.

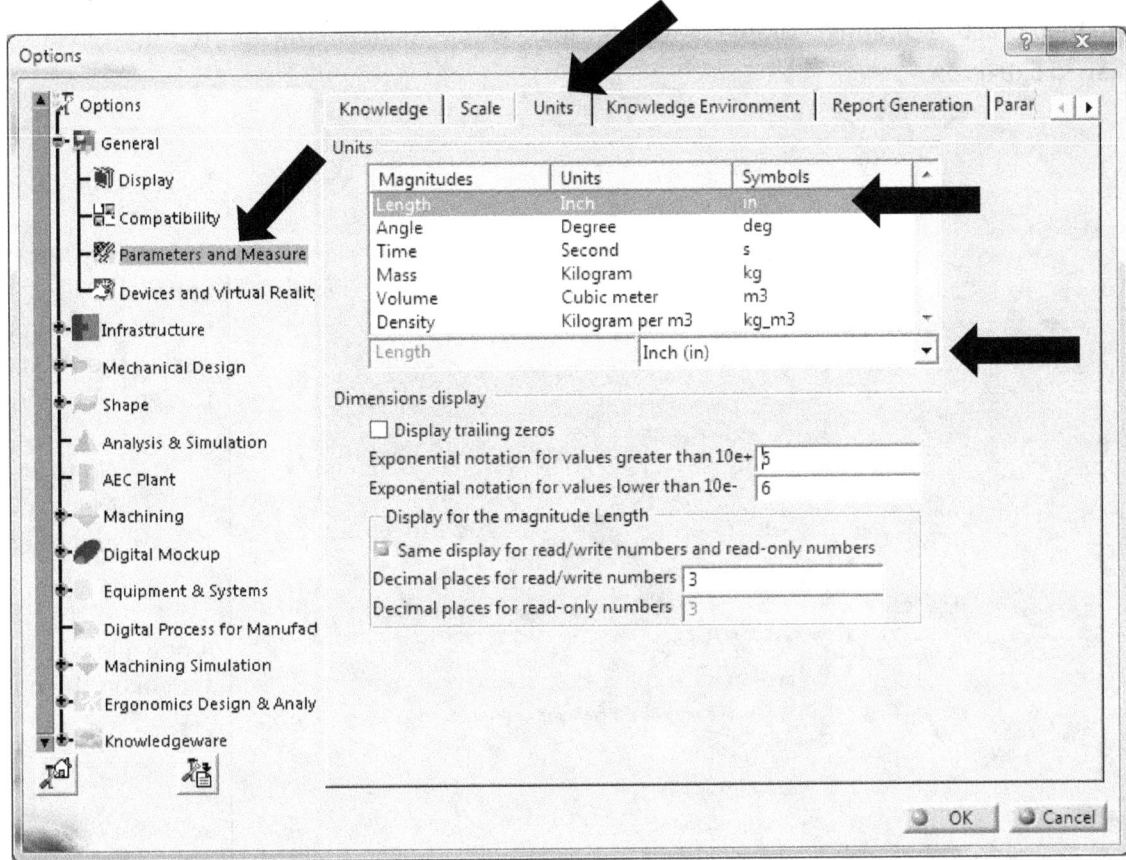

In order to change the setups on each workbench, select an appropriate Application Tool on the left side.

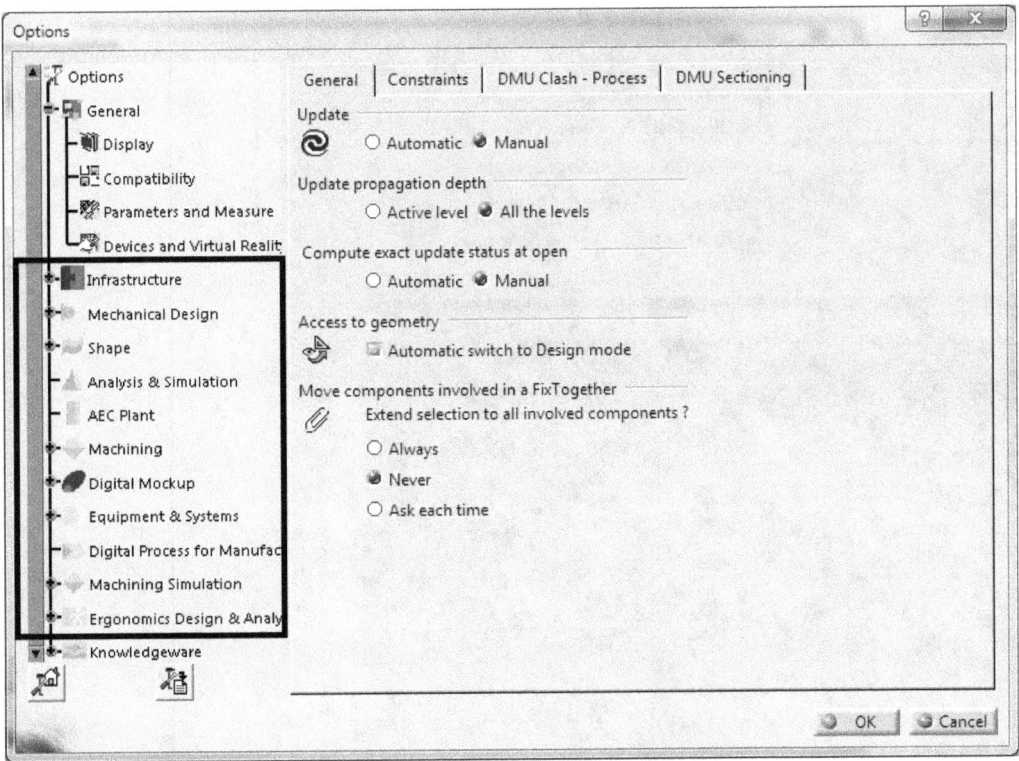

Chapter 2 – Sketcher Workbench

This chapter is to help practice creating a simple sketch, finishing with a solid. Follow the step-by-step instructions below to create a T-shaped Extrusion.

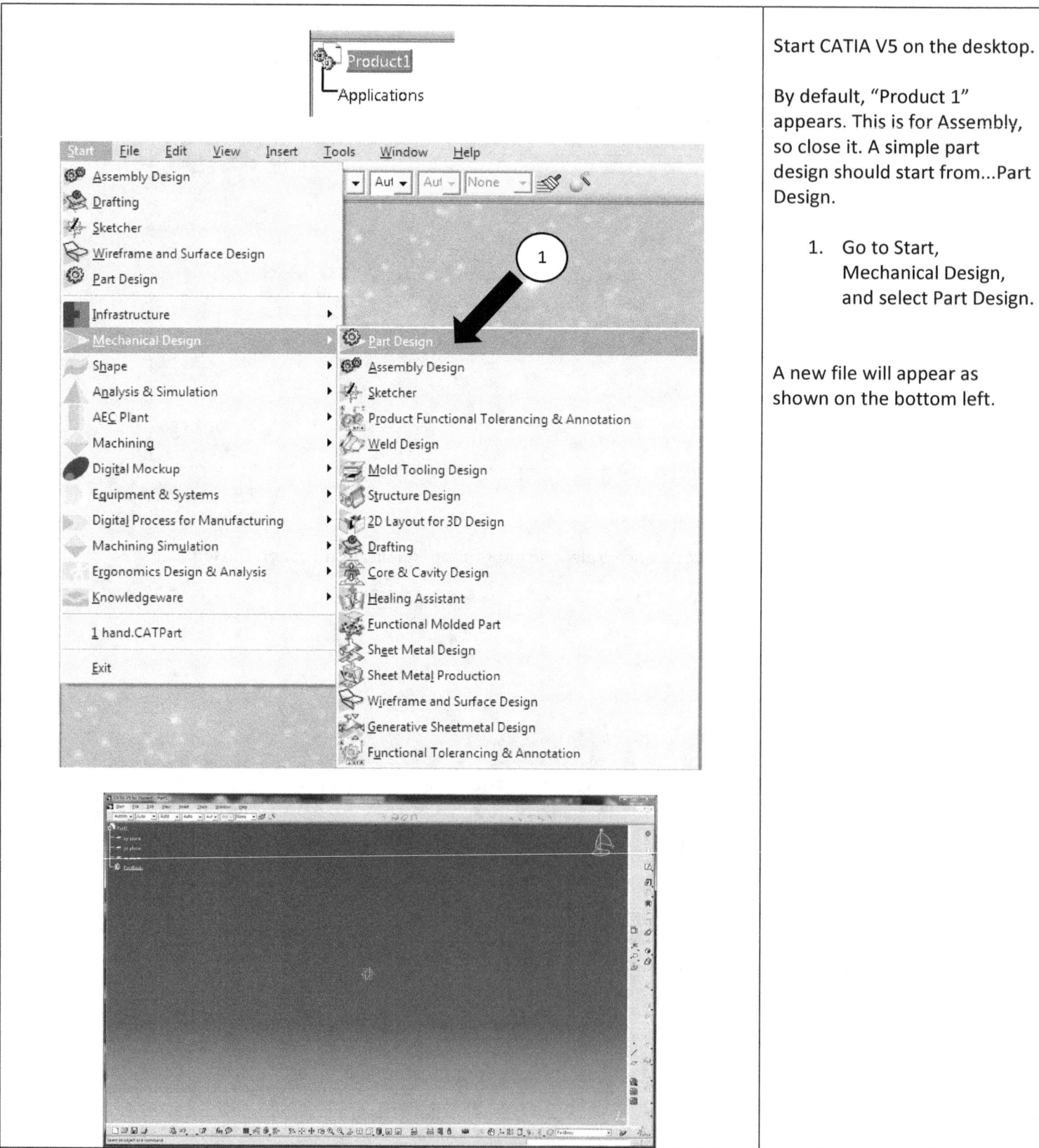

Start CATIA V5 on the desktop.

By default, "Product 1" appears. This is for Assembly, so close it. A simple part design should start from...Part Design.

1. Go to Start, Mechanical Design, and select Part Design.

A new file will appear as shown on the bottom left.

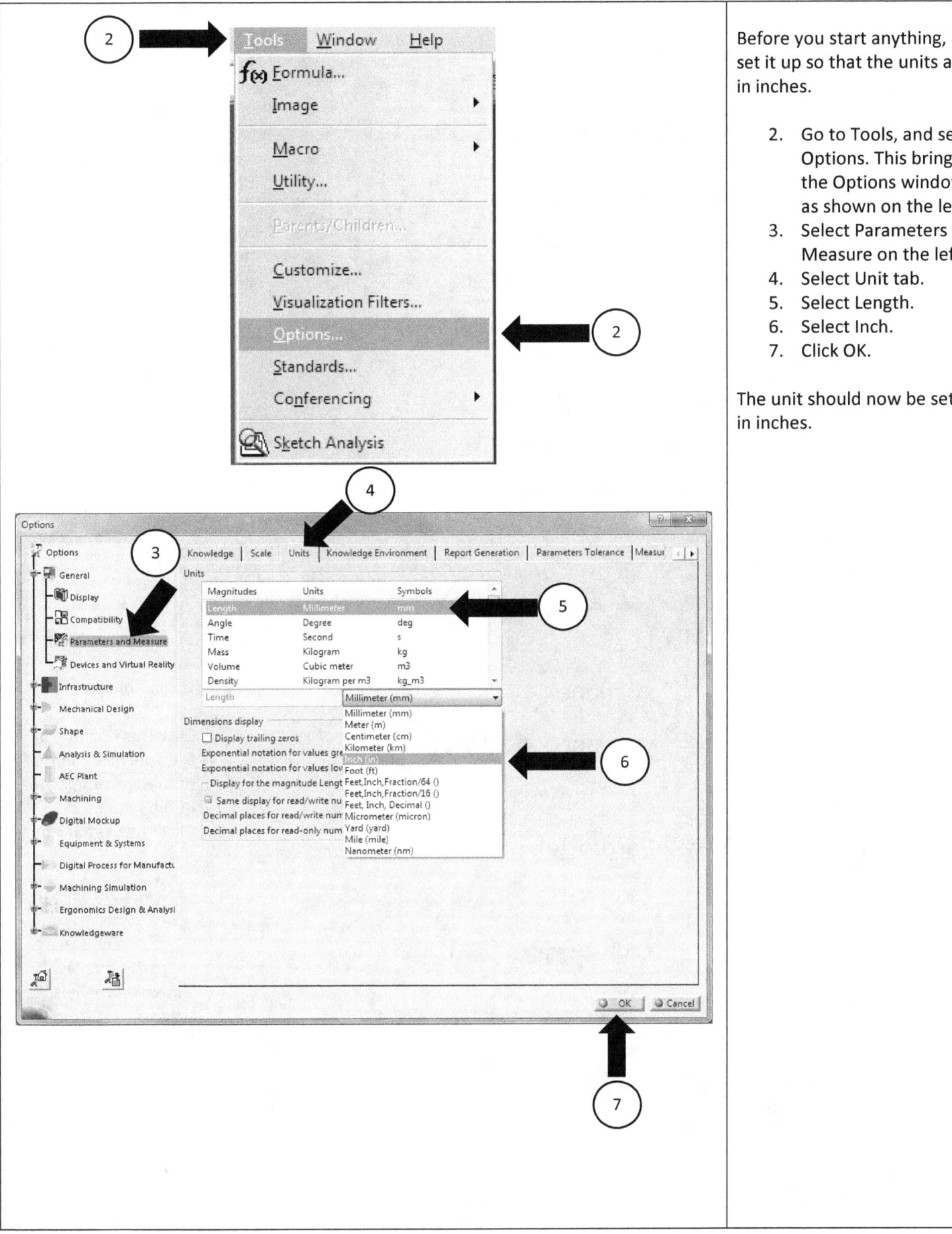

Before you start anything, let's set it up so that the units are in inches.

2. Go to Tools, and select Options. This brings up the Options window as shown on the left.
3. Select Parameters and Measure on the left.
4. Select Unit tab.
5. Select Length.
6. Select Inch.
7. Click OK.

The unit should now be set up in inches.

Profile: continuous sketching

Rectangle: to create a rectangular shape

Circle: to create a circular/arc shape

Spline: to create undefined curvature shape

Ellipse: to create an ellipse

Line: to create straight lines

Axis: to create an axis line

Point: to create points

Start sketching in inches. To do that, you must have the correct toolbar--"Profile" as shown on the left.

A single click on an icon means it activates only once.

Double clicking on an icon means it remains activated until you click the icon again.

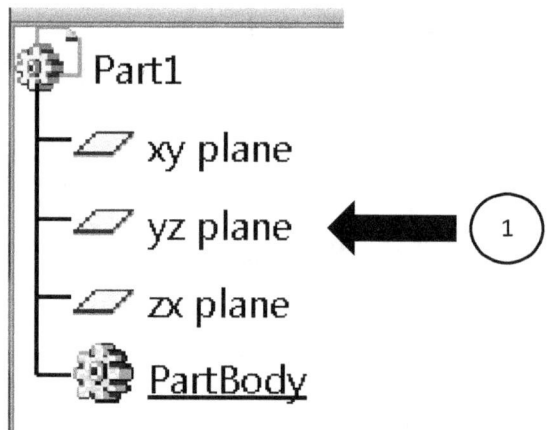

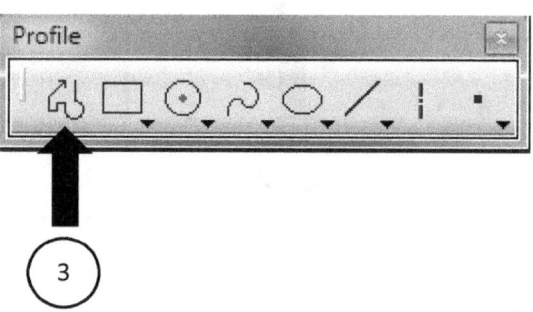

Let's start sketching.

1. Click the YZ plane on the Specification tree or in the Graphic area in the middle.
2. Click the Sketch icon
3. Use the Profile icon and create a t-shaped sketch as shown on the left. Be sure to start from the Origin.

12

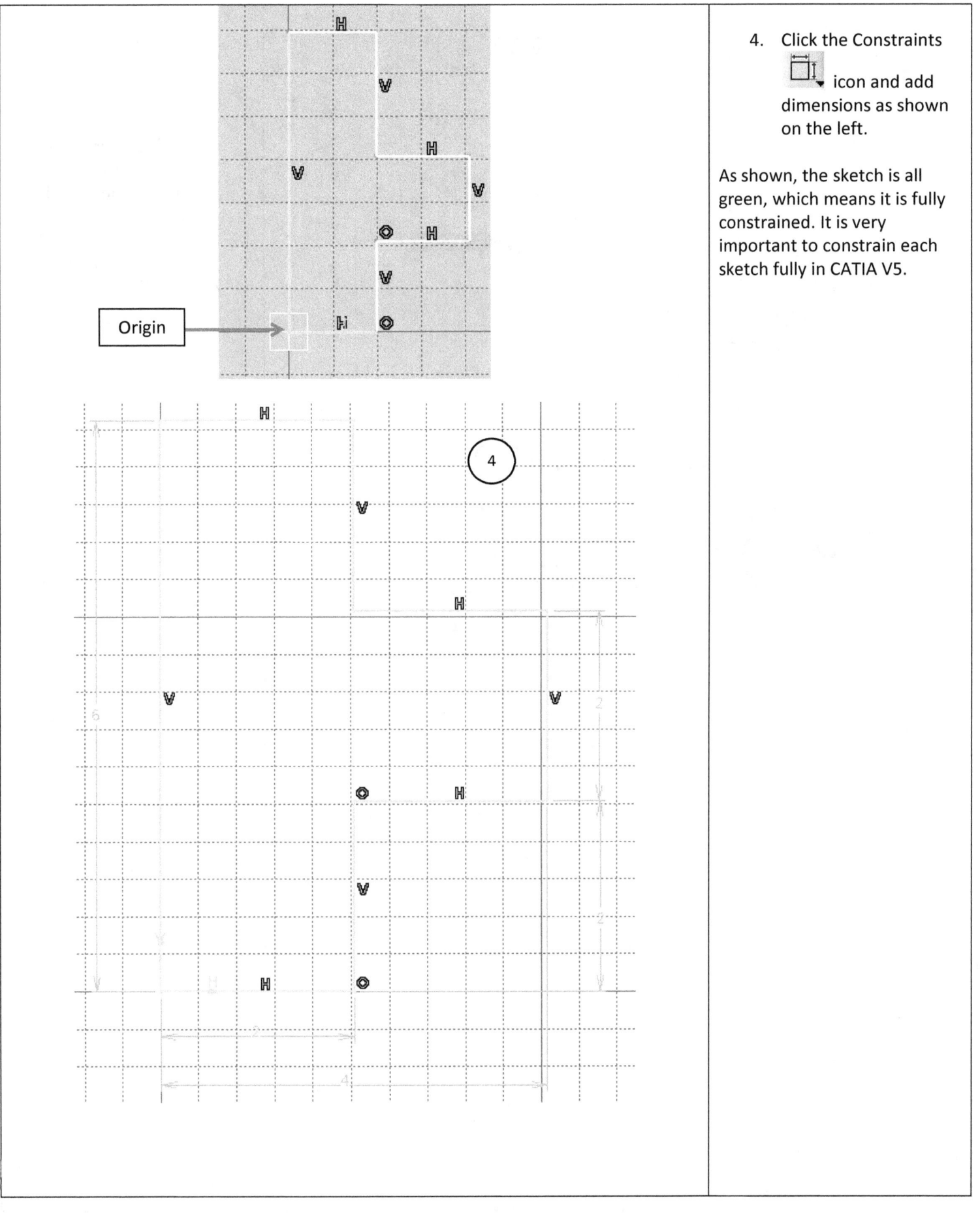

4. Click the Constraints icon and add dimensions as shown on the left.

As shown, the sketch is all green, which means it is fully constrained. It is very important to constrain each sketch fully in CATIA V5.

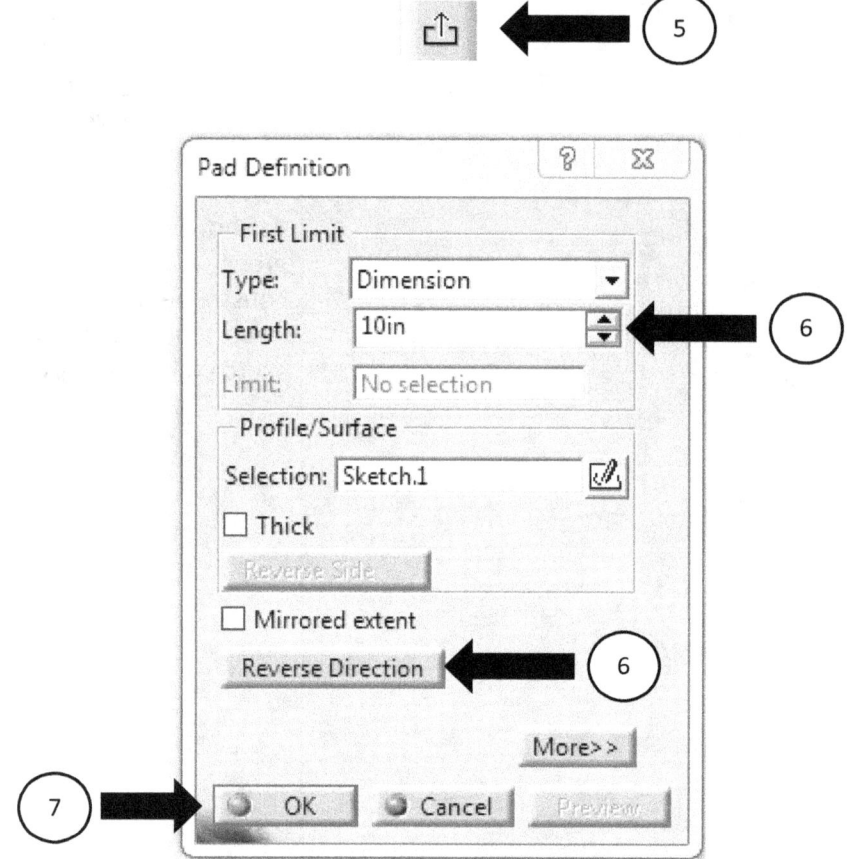

5. Click on the Exit workbench icon
6. Click on the Pad icon and enter 10 in the "Length" slot. Change the direction of pad by clicking "Reverse Direction" in the Pad Definition box.
7. Click OK.

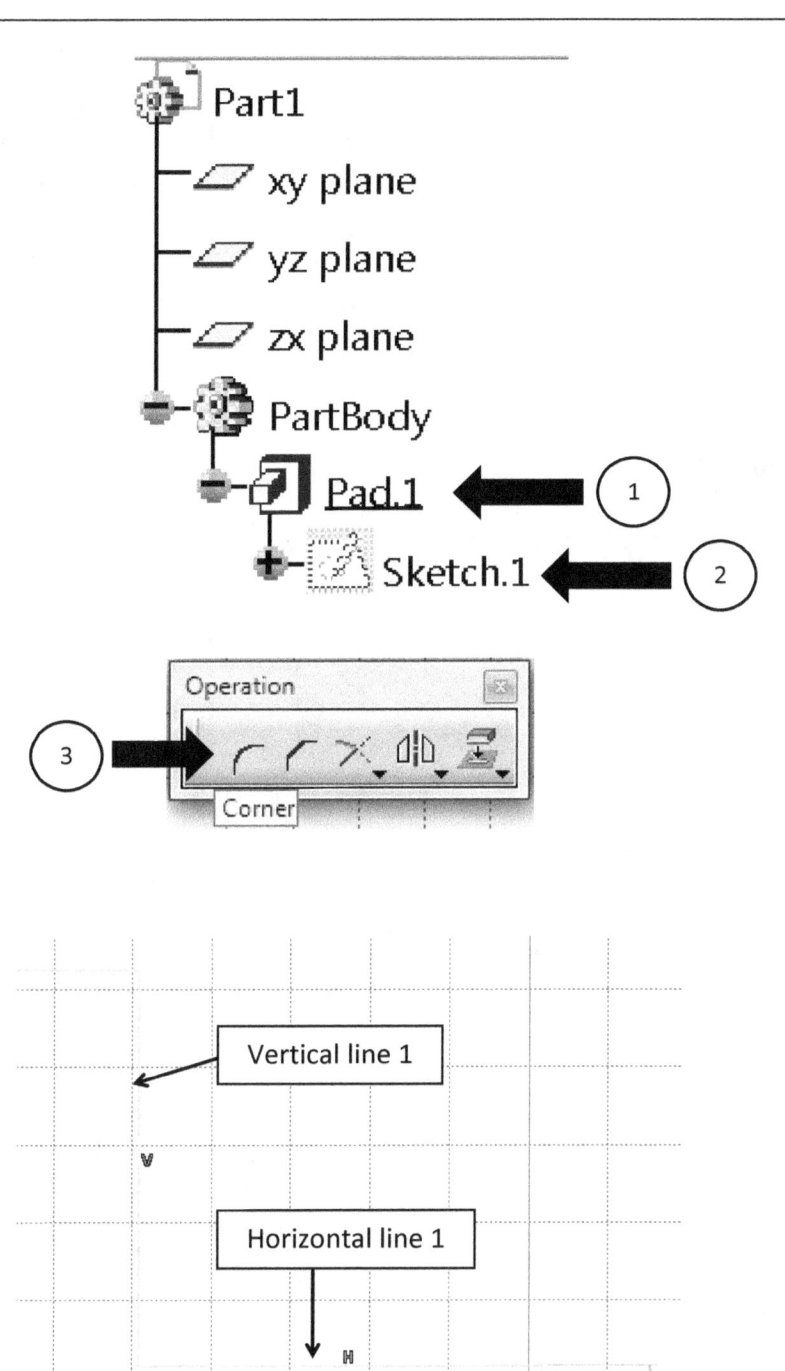

Let's edit the sketch.

1. Click on the plus sign of Pad.1. on the tree and show Sketch.1
2. Double click on Sketch.1. This will take you to the Sketcher Workbench.
3. Click on the Corner icon on the Operation toolbar as shown on the left.
4. Select Vertical line 1 and then Horizontal line 1 as shown on the left. Then left-click.

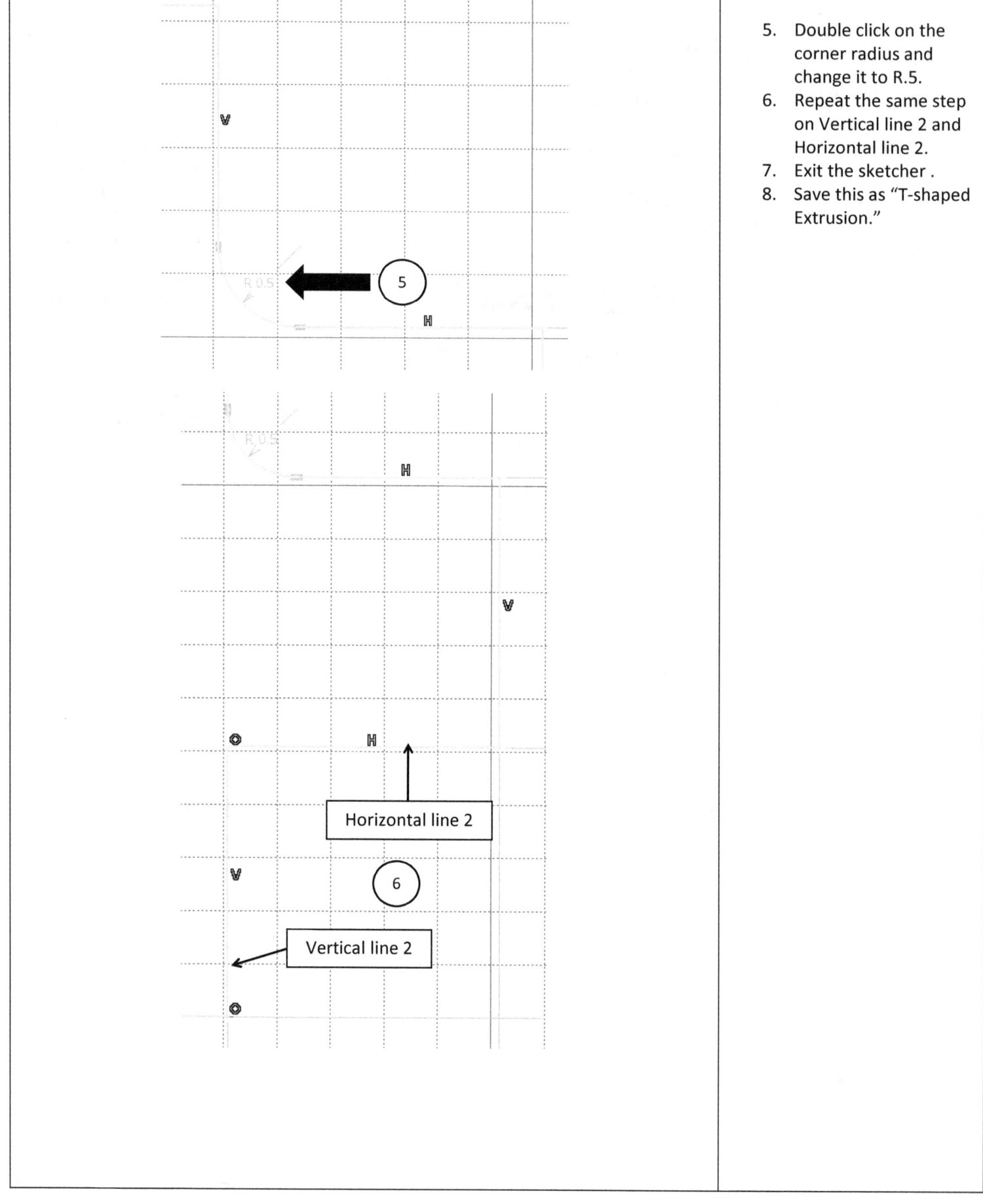

5. Double click on the corner radius and change it to R.5.
6. Repeat the same step on Vertical line 2 and Horizontal line 2.
7. Exit the sketcher.
8. Save this as "T-shaped Extrusion."

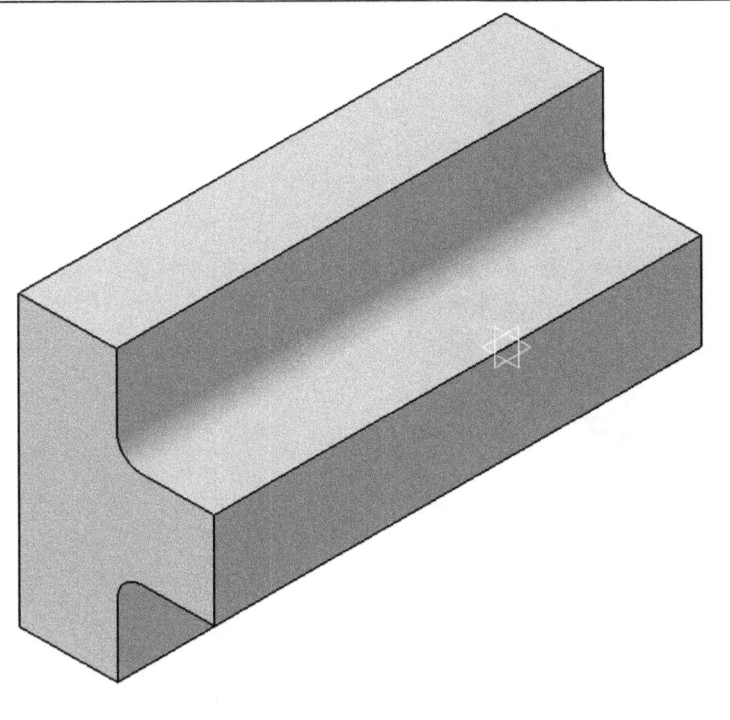

The part now has fillets as shown on the left.

You can also add fillets to a three-dimensional solid.

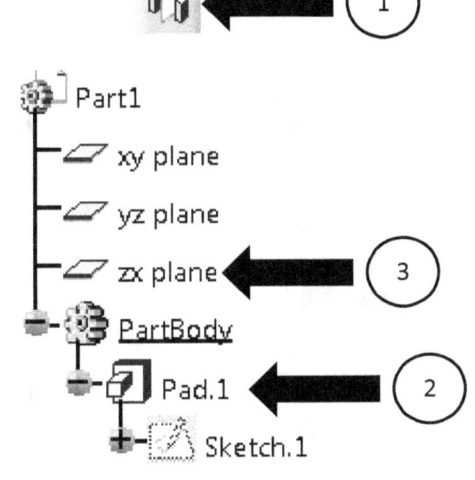

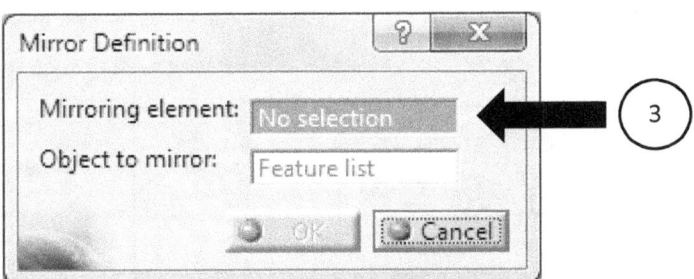

Now let's mirror this part.

1. Click on the mirror icon
2. Select the Pad. 1 from the Tree.
3. Select the ZX plane from the Tree for the Mirroring element.
4. Click OK.

17

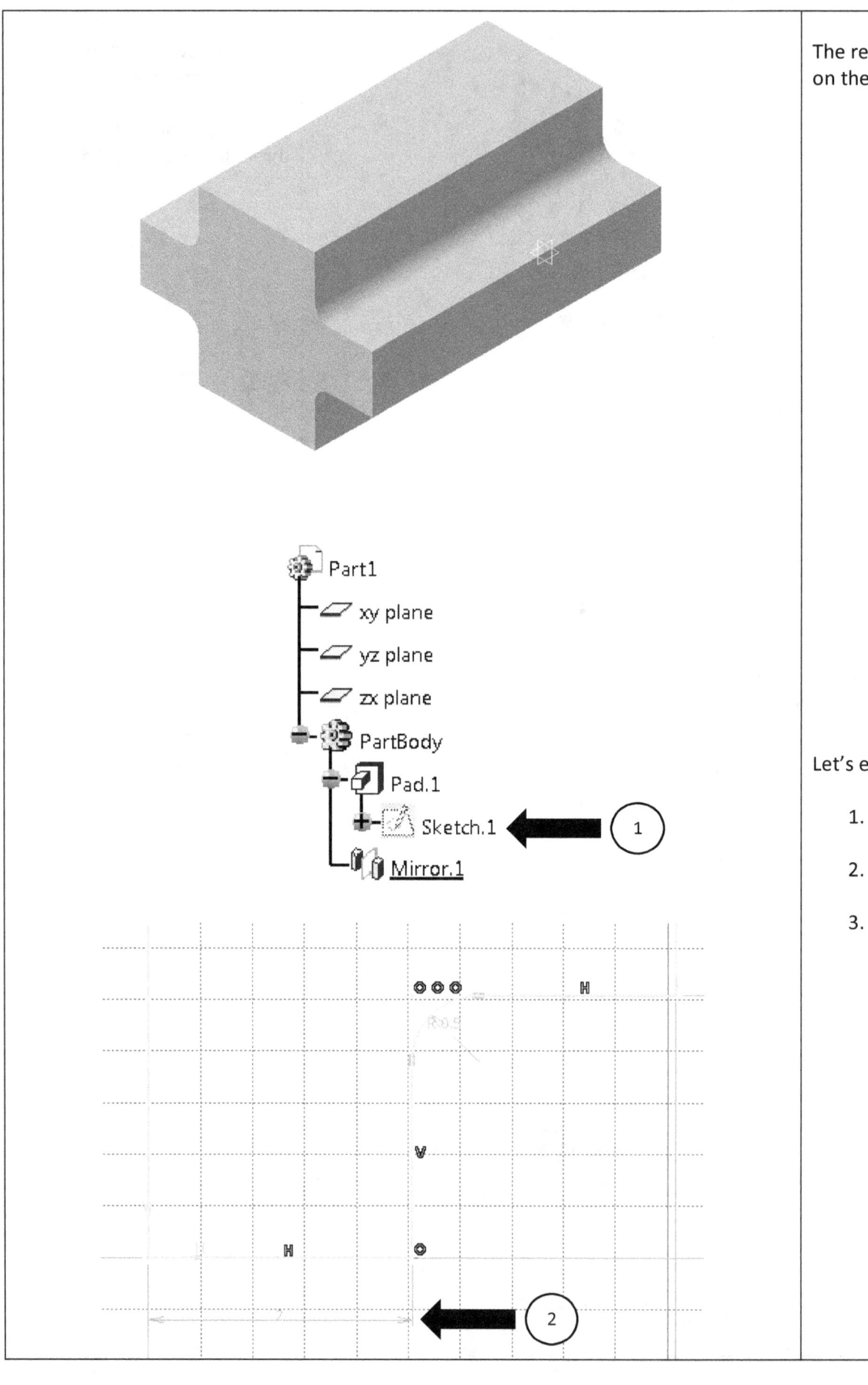

The result should look like this on the left.

Let's edit the sketch.

1. Double-click on the Sketch.1 on the Tree.
2. Change the horizontal value from 2 to 1.
3. Exit Sketcher.

18

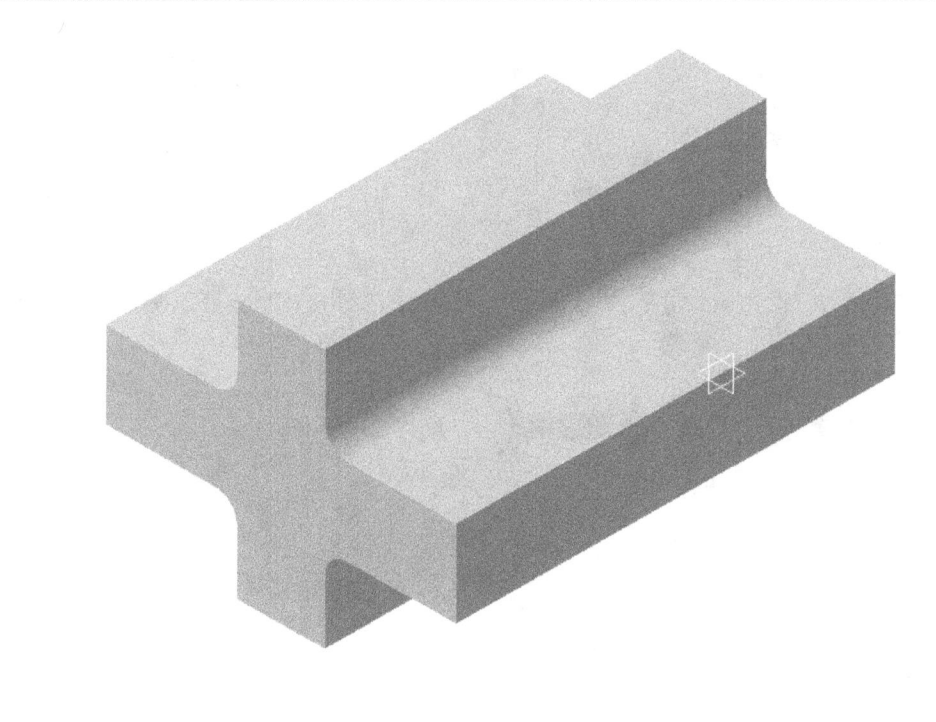

4. Observe the result. It should be similar to the sketch on the left.
5. Save as "Cross".

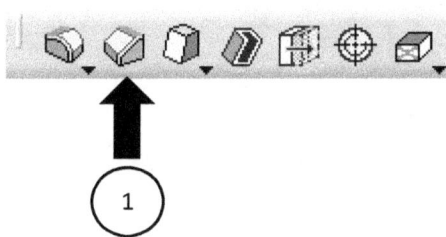

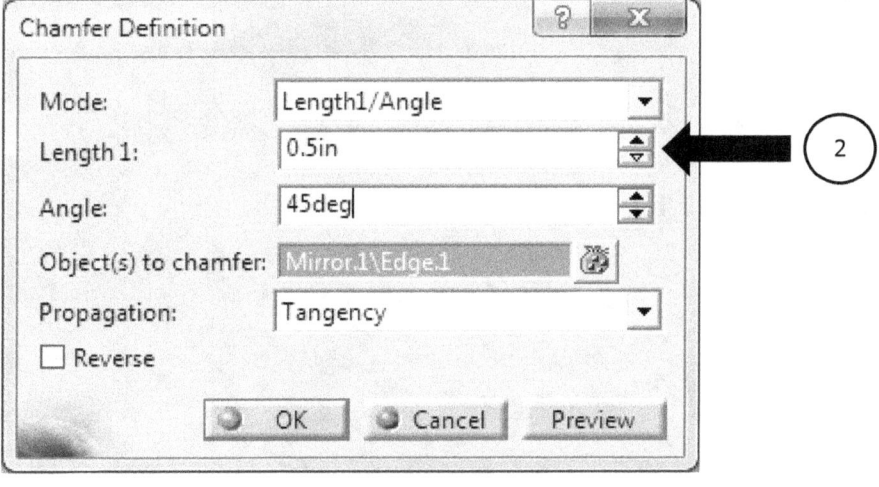

Keep "Cross" opened from the previous chapter.

Let's add chamfers.

1. Click on the Chamfer icon
2. Enter 0.5 in Length 1 within the Chamfer Definition box.

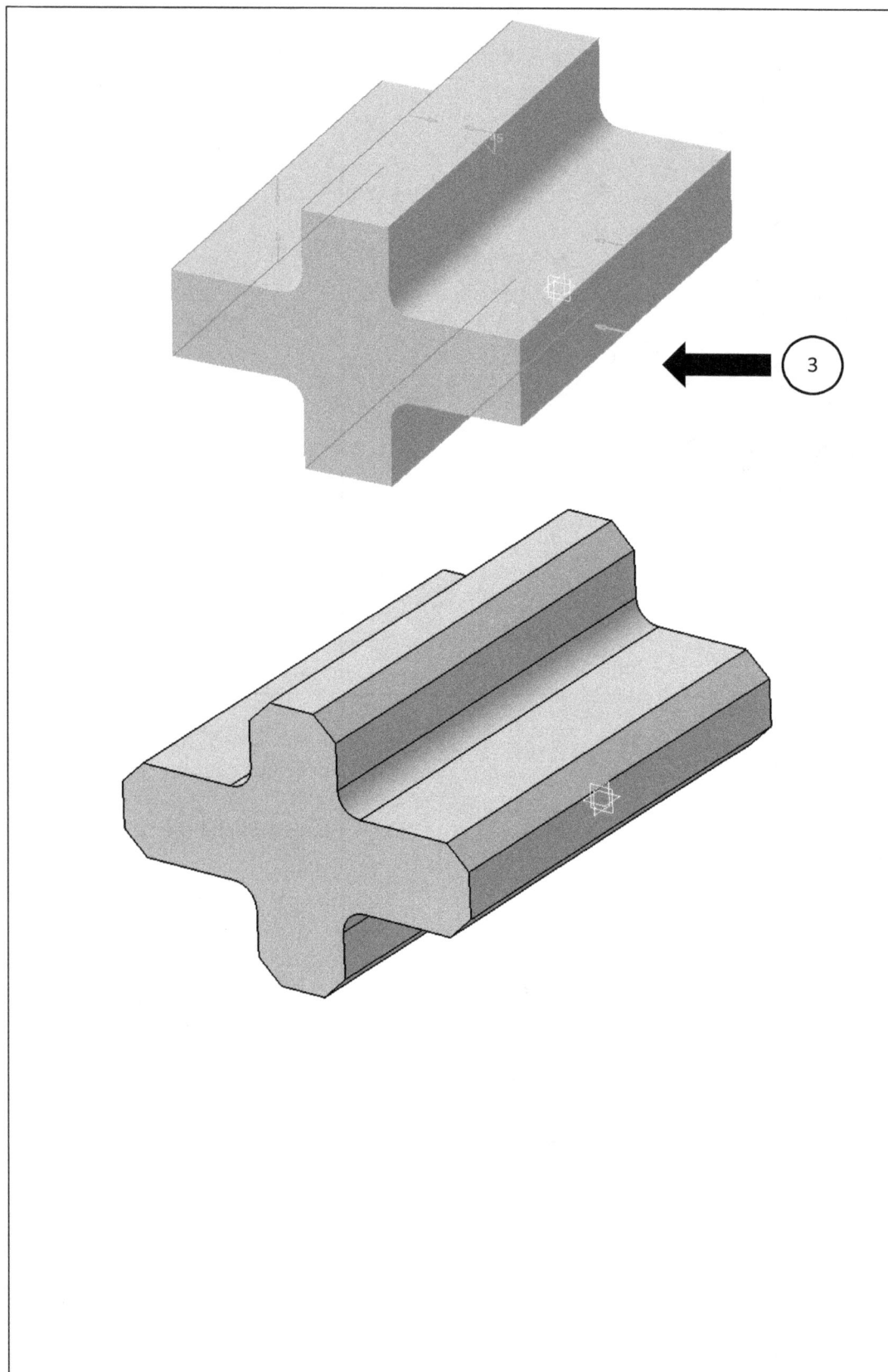

3. Select each edge of the part as shown on the left.
4. Click OK.
5. Save as "Chamfer."

The result should appear as shown on the left.

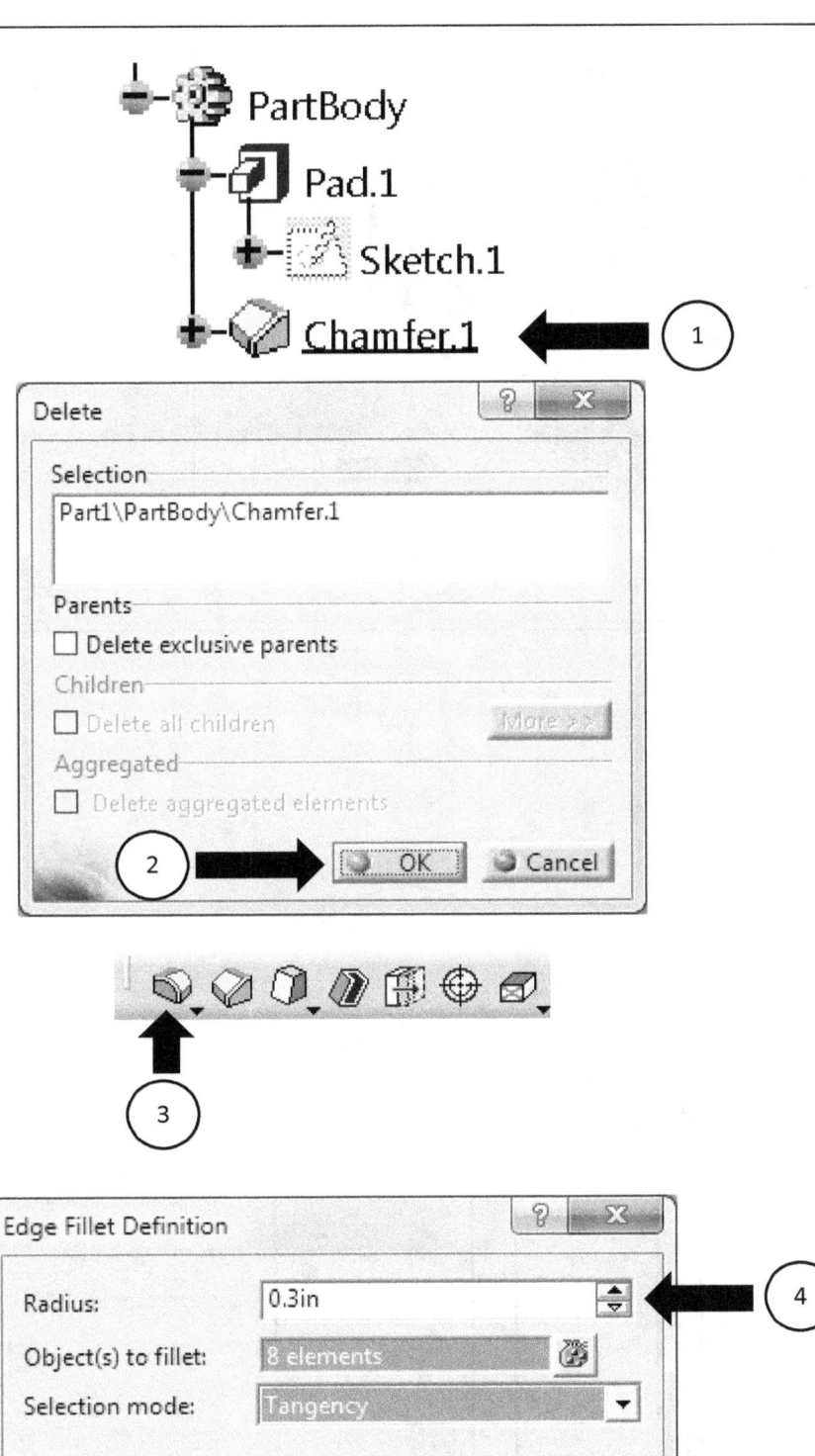

Let's change the Chamfers to Fillets.

1. Delete Chamfer.1 from the Tree. (Right click and select Delete, or hit Delete key on the keyboard.)
2. Delete box appears, click OK.
3. Click on the Edge Fillet icon.
4. Enter 0.3 in Radius in the Edge Fillet Definition box.

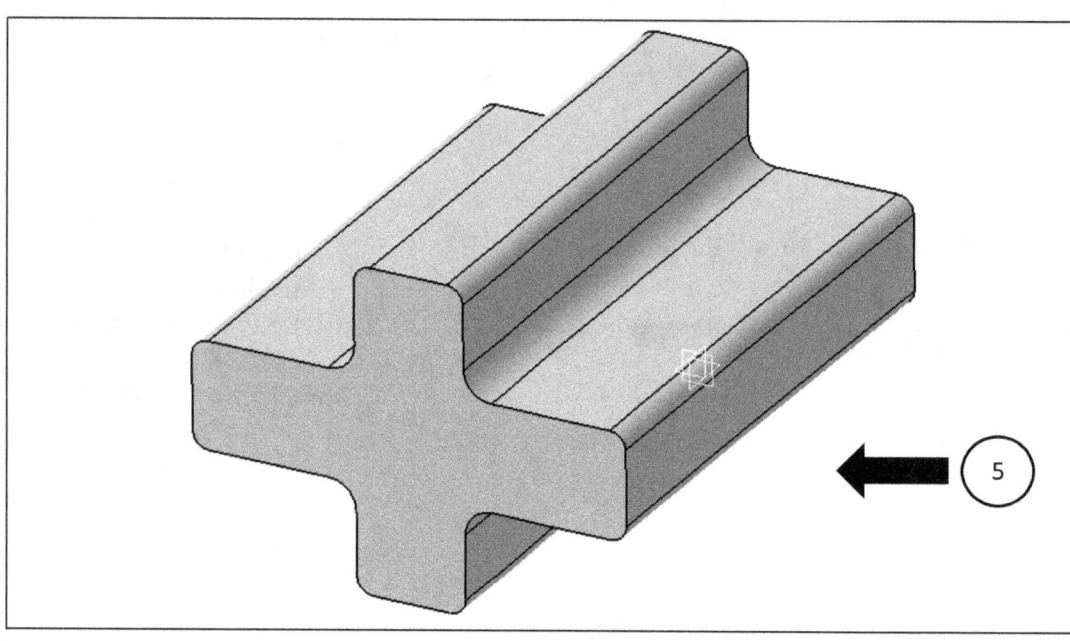

5. Click on every edge on the part.
6. Save as "Fillet."

The result should appear as shown on the left.

Chapter 2 Assignment

1: Unit: inches

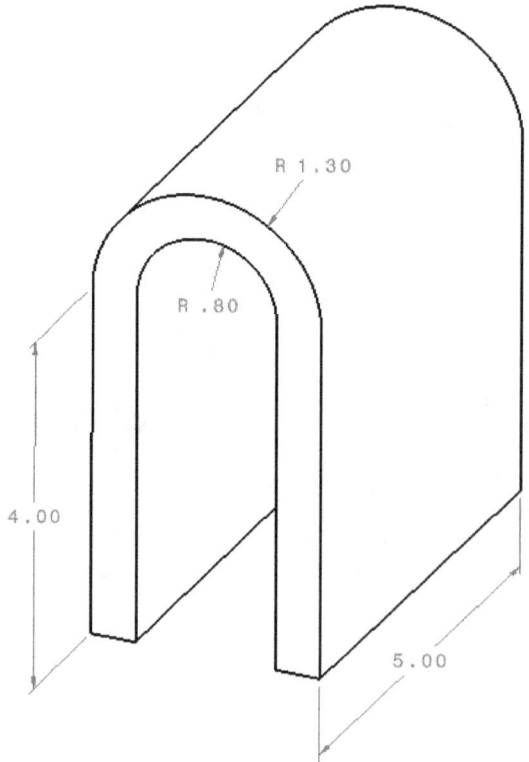

2: Unit: mm. Size and scale are up to you. Add thickness of 50mm.

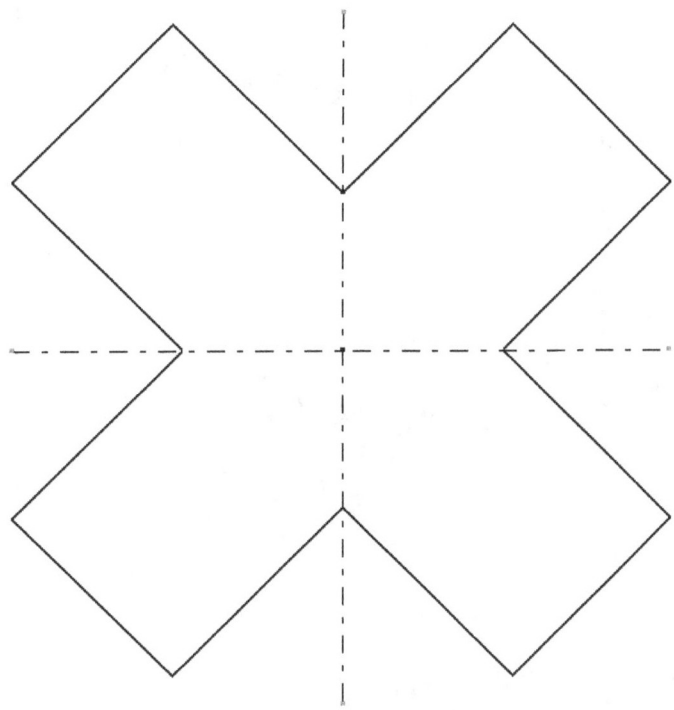

3: Unit: inches. Size and scale are up to you

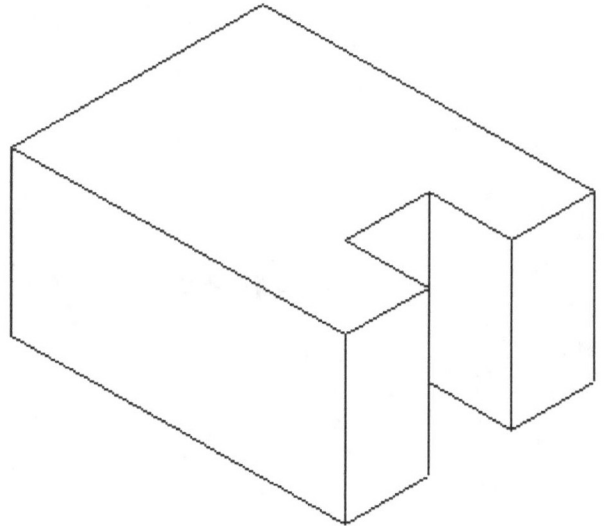

Chapter 3 – Part Design Workbench

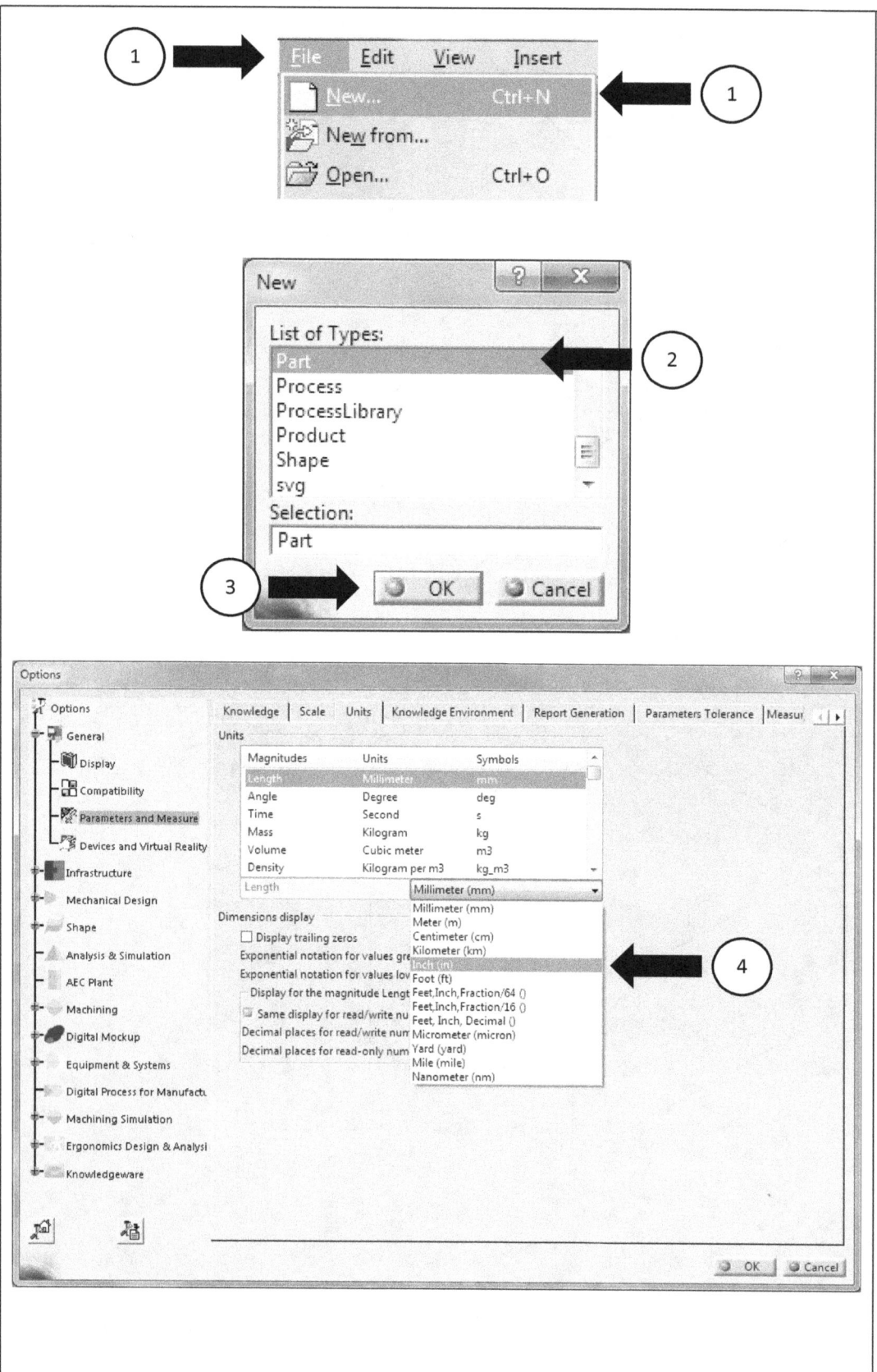

1. Go to File and select New.
2. Select "Part" from the List of Types.
3. Click OK.
4. Go to Tools and select Options to set the parameters to mm.

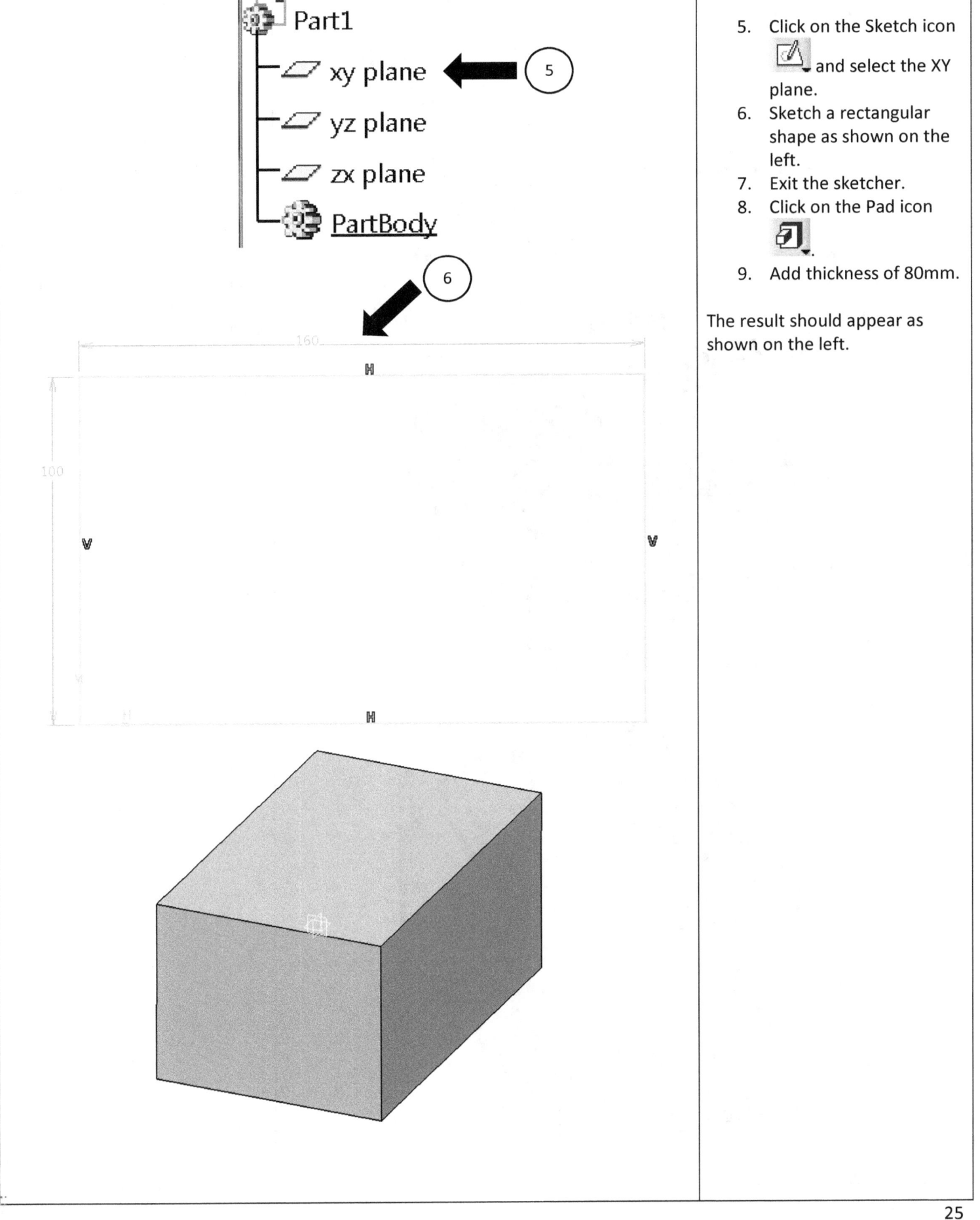

5. Click on the Sketch icon ✎ and select the XY plane.
6. Sketch a rectangular shape as shown on the left.
7. Exit the sketcher.
8. Click on the Pad icon.
9. Add thickness of 80mm.

The result should appear as shown on the left.

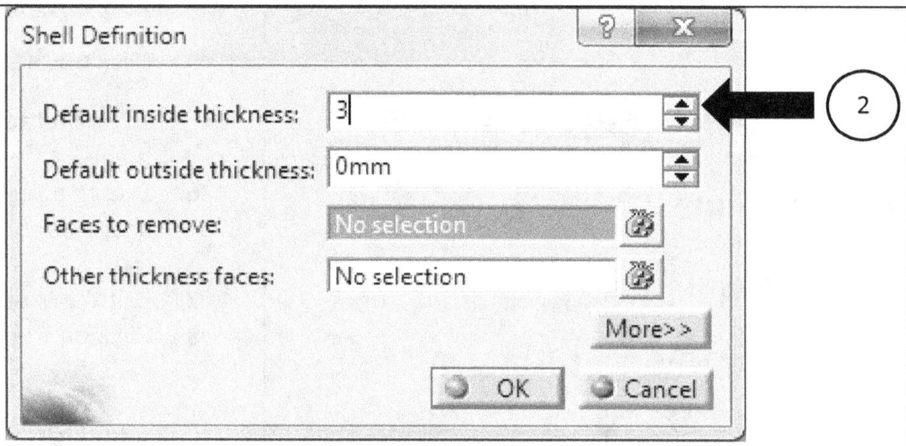

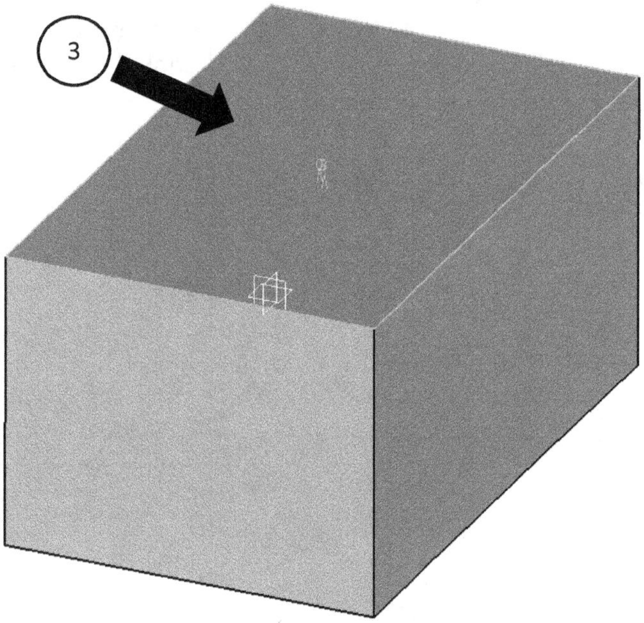

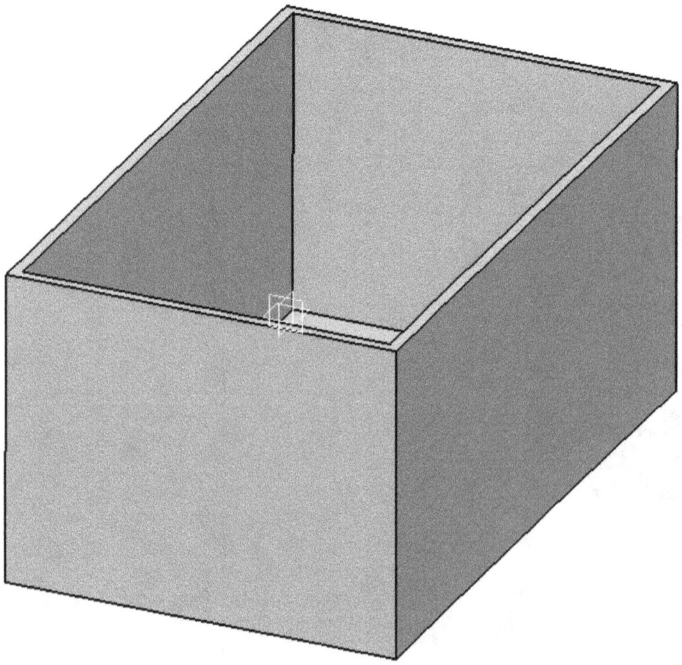

Let's practice Shell icon.

1. Click on the Shell icon .
2. Enter 3 in Default inside thickness in the Shell Definition box.
3. Click on the top surface on the part.
4. Click OK.

It should look like on the left. Now it has a wall 3 millimeters thick.

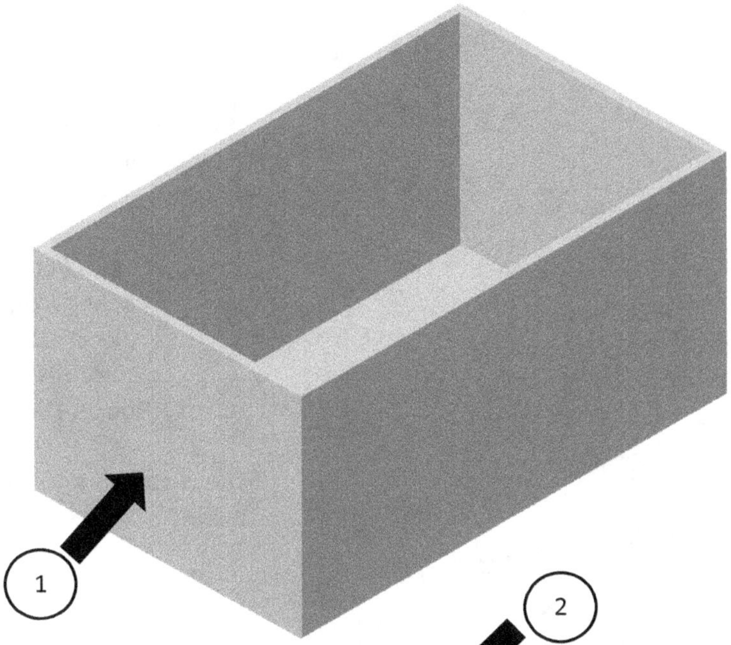

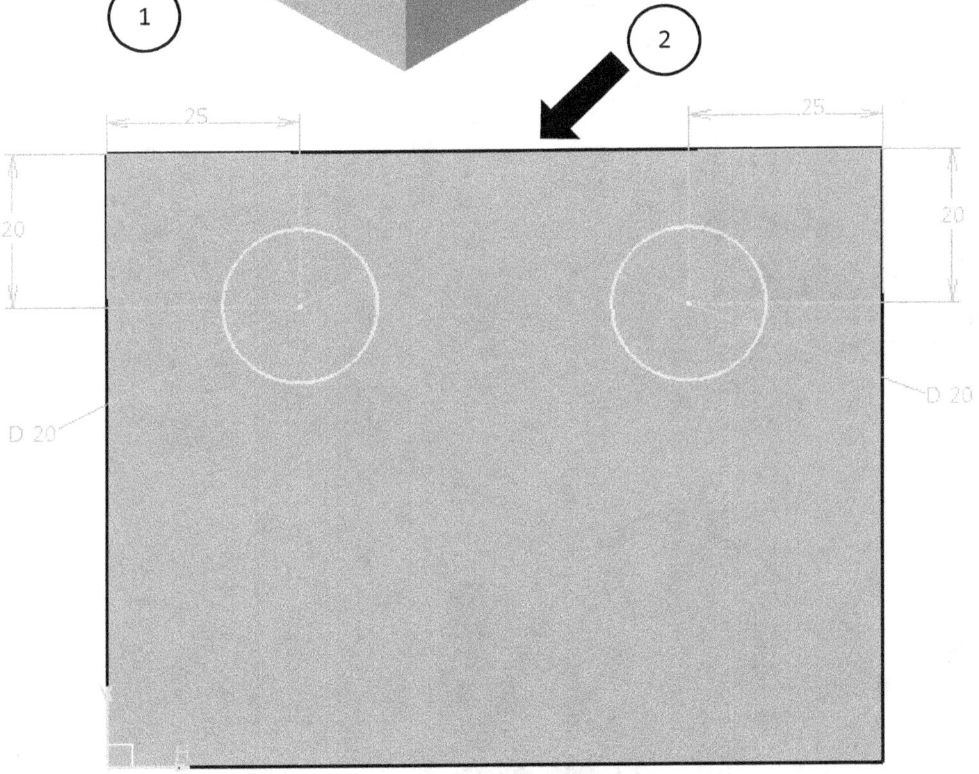

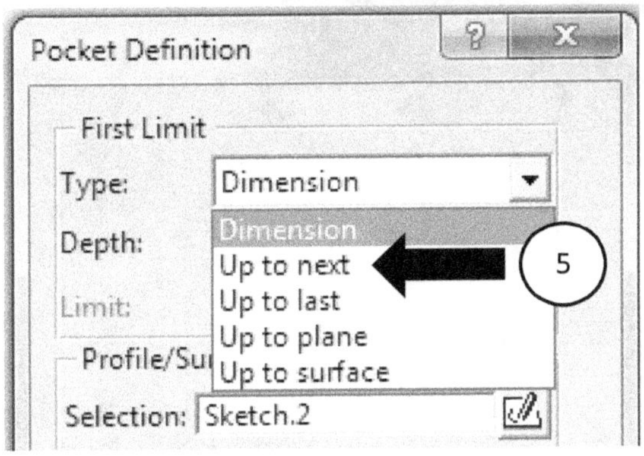

Let's add holes. There are two ways to create holes. One is to start with a sketch, and the other is by using the Hole icon .

Let's start with a sketch.

1. Select the front surface (100 x 80).
2. Create a sketch as shown on the left.
3. Exit the sketcher.
4. Click on the Pocket icon .
5. Select "Up to next" under the Type in the Pocket Definition box.
6. Click OK.

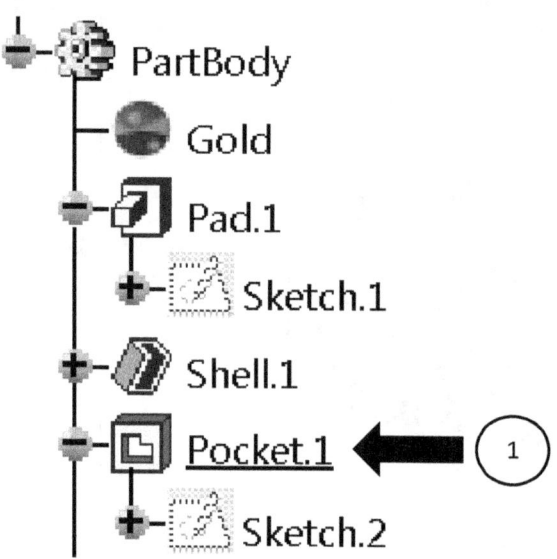

Two holes were created on the front wall.
Let's edit these holes so that they go all the way through the box.

1. Double-click on the Pocket.1 on the Tree.
2. Select "Up to last" under the Type in the Pocket Definition box.

The result should be similar to the left.

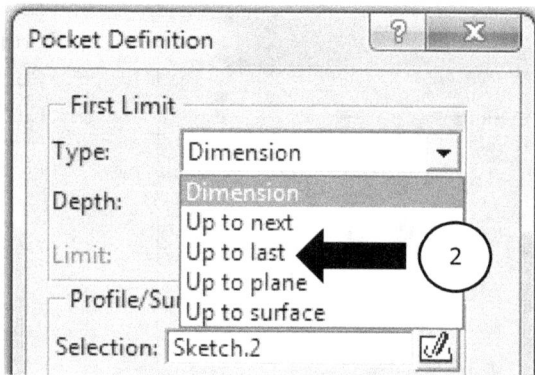

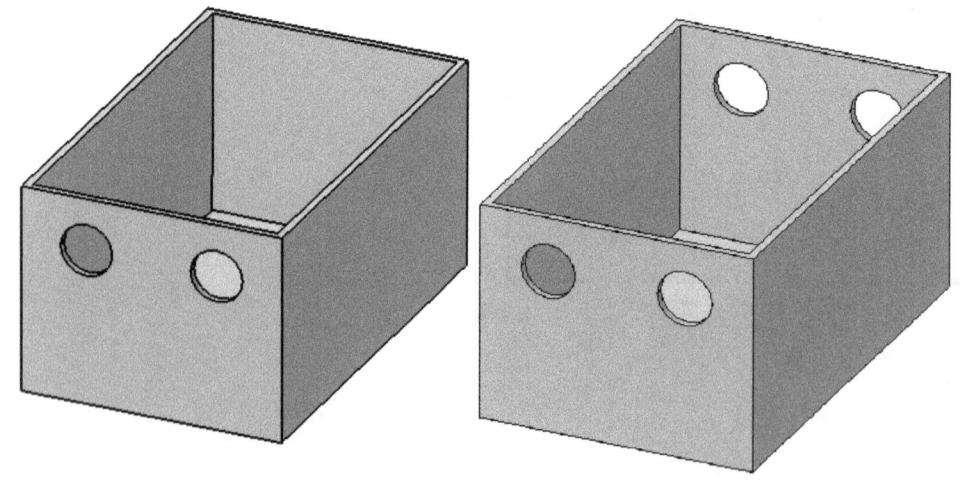

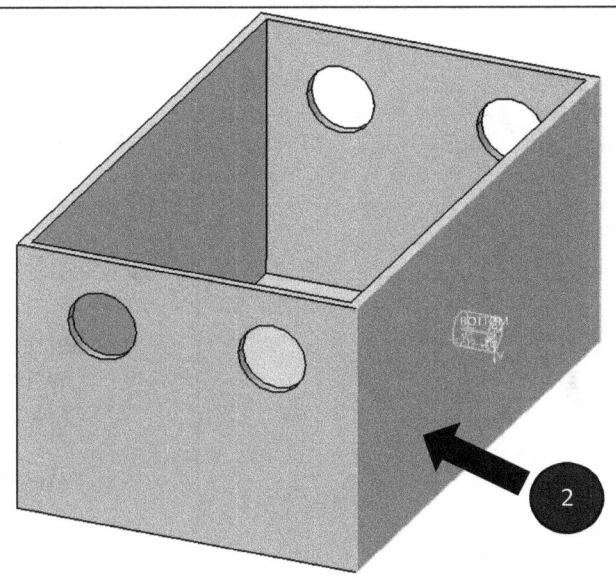

Let's practice using the Hole icon .

1. Click on the Hole icon .
2. Click on the right side surface of the part.
3. Select "Up To Last"
4. Enter 20 in the Diameter.
5. Then click on the Sketch icon in the Hole Definition box – this is to locate the hole.
6. Use the constraint icon , locate the center of the hole as shown on the left.
7. Exit the Sketcher.
8. Click OK.

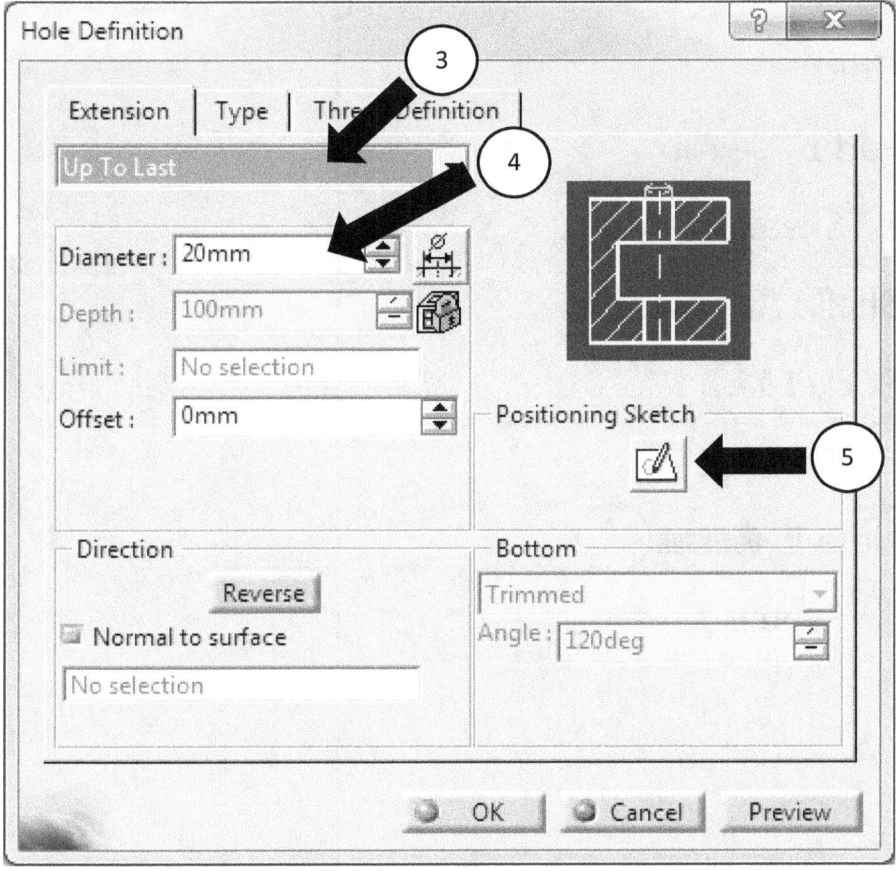

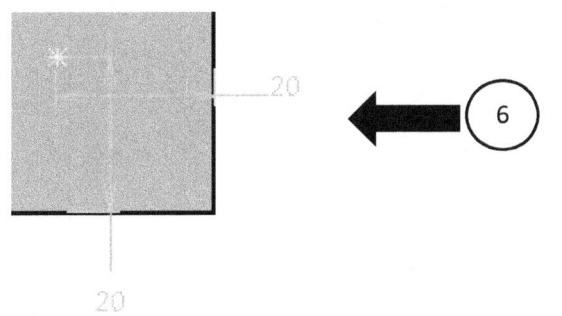

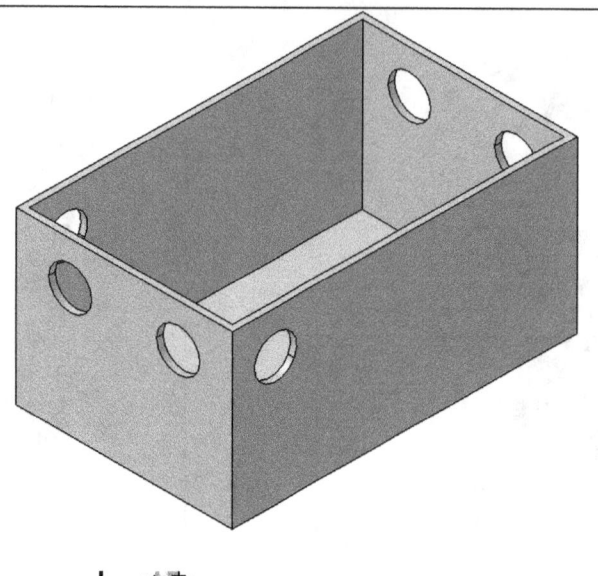

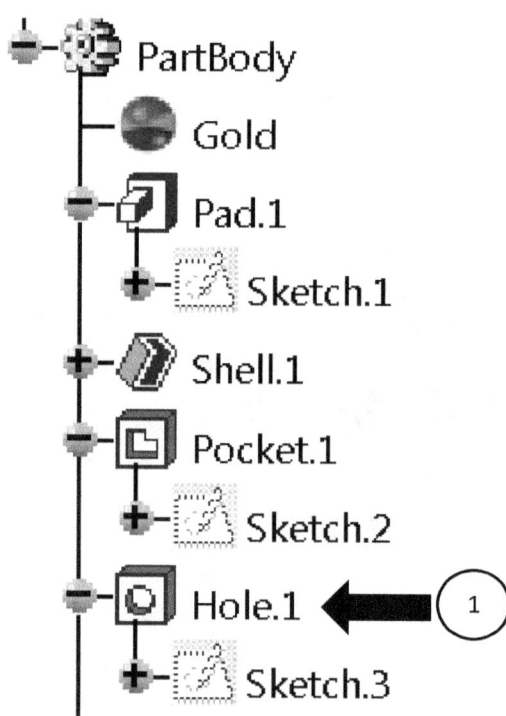

The result should appear similar to the box on the left.

Let's practice the Pattern icon.

1. Select the Hole.1 from the Tree.
2. Click the Rectangular Pattern icon.

30

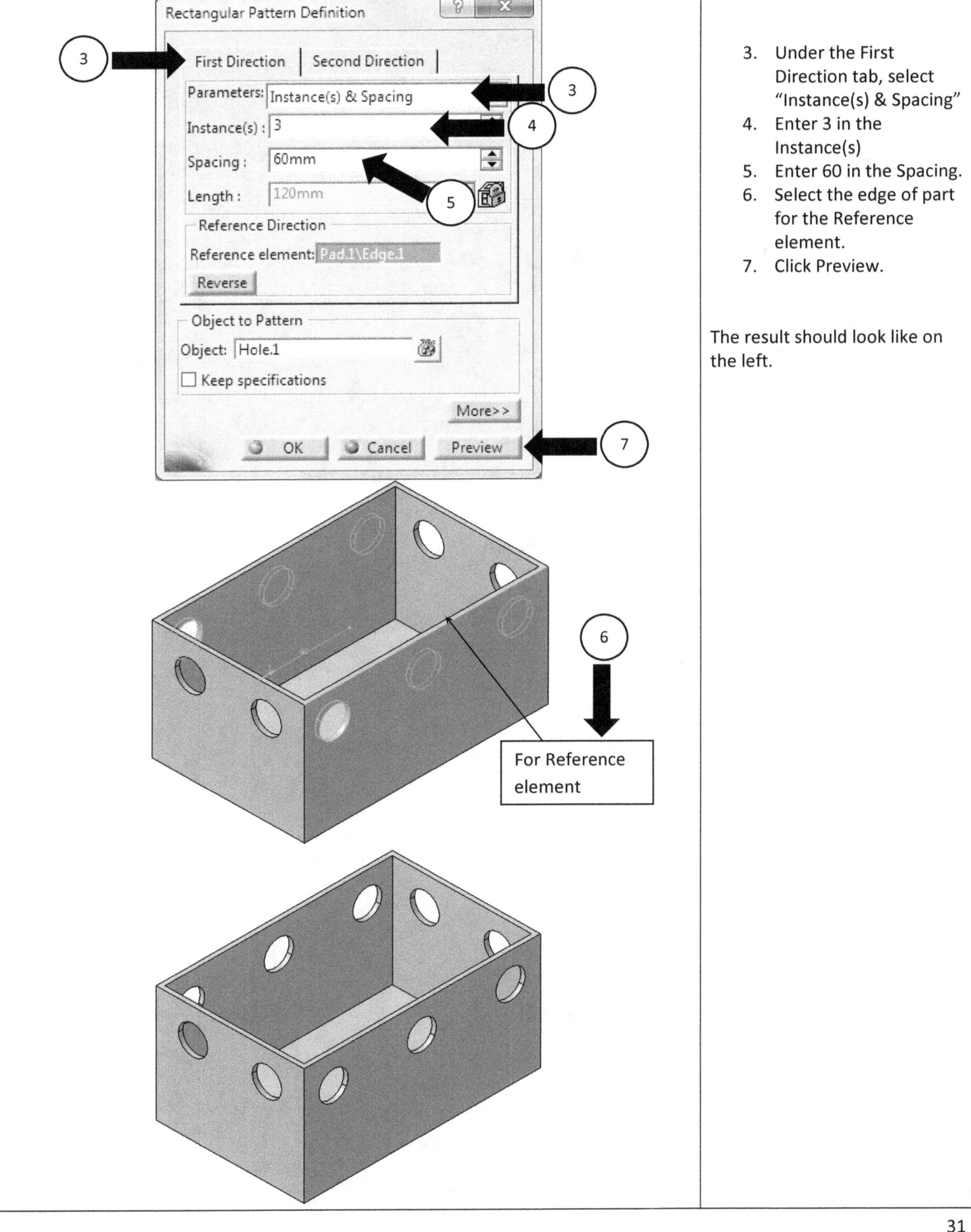

3. Under the First Direction tab, select "Instance(s) & Spacing"
4. Enter 3 in the Instance(s)
5. Enter 60 in the Spacing.
6. Select the edge of part for the Reference element.
7. Click Preview.

The result should look like on the left.

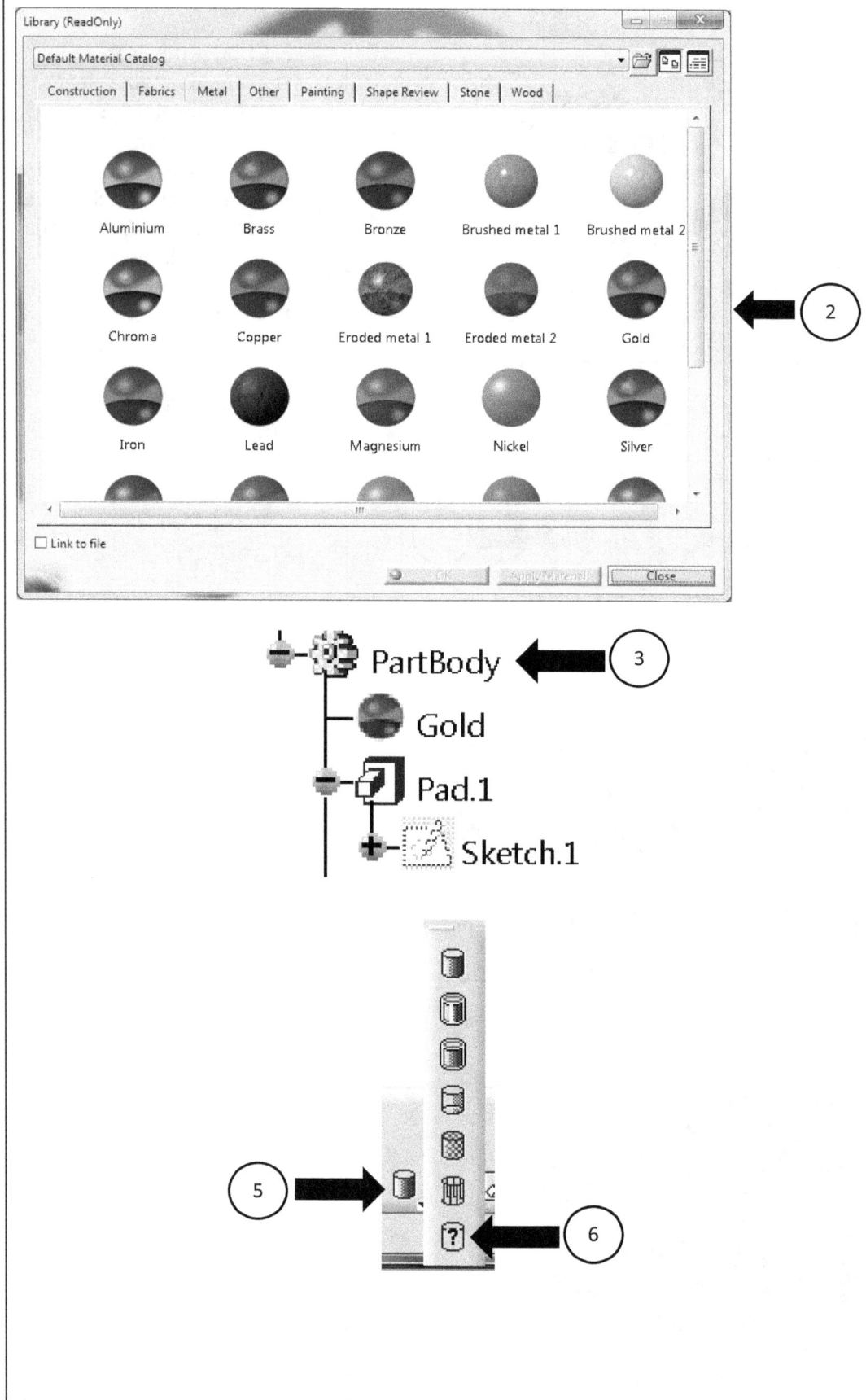

Let's apply material.

1. Click on the Apply Material icon.
2. Select a material of your choice.
3. Click PartBody on the Tree.
4. Click OK.
5. Click the arrow in the Shading icon.
6. Click the Customize View Parameters.

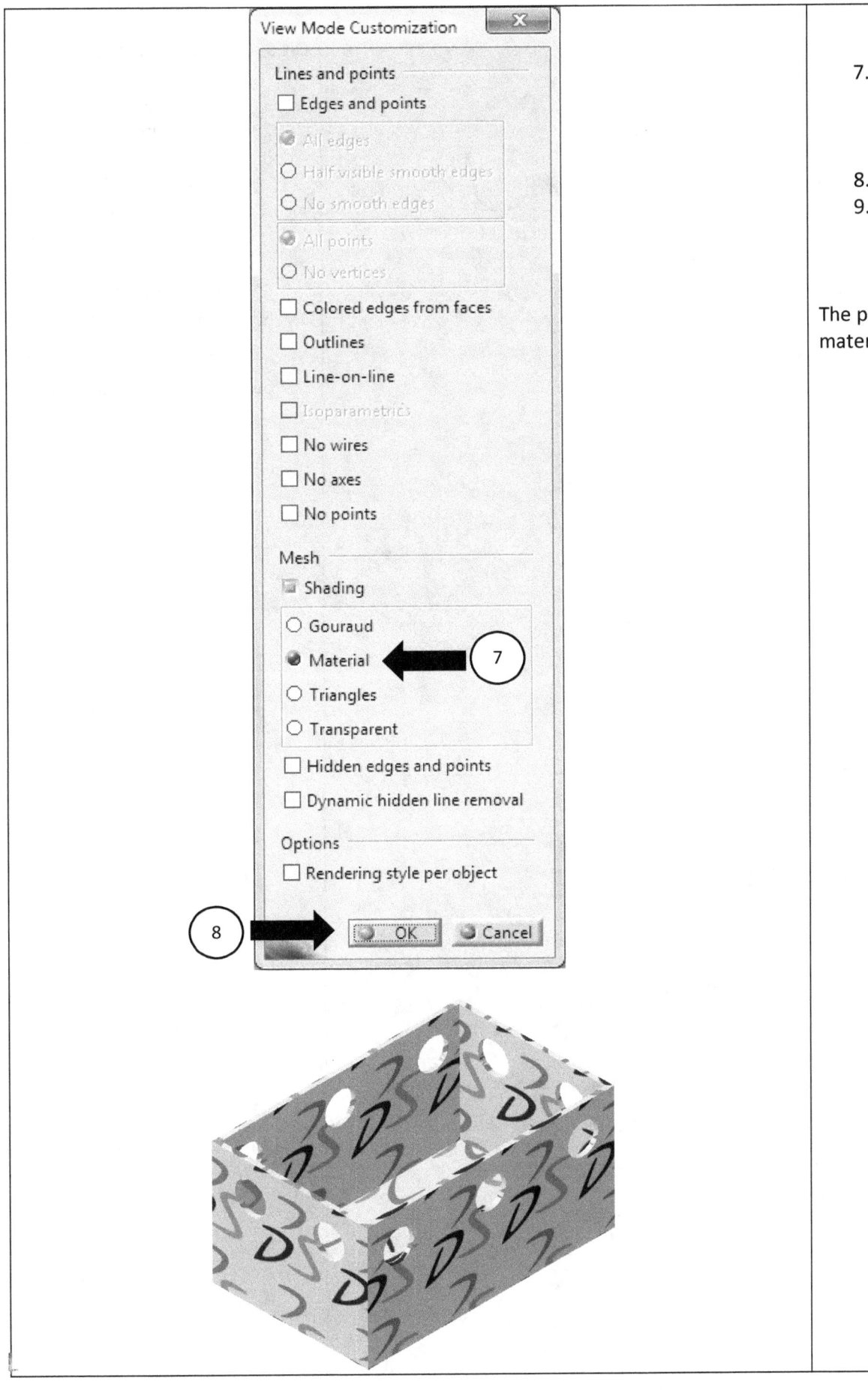

7. The View Mode Customization will show up, then select Material.
8. Click OK.
9. Save as "Shaded."

The part should appear with the material of your choice.

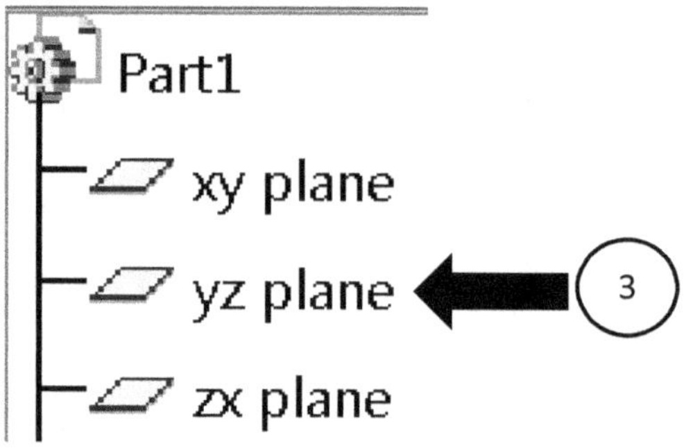

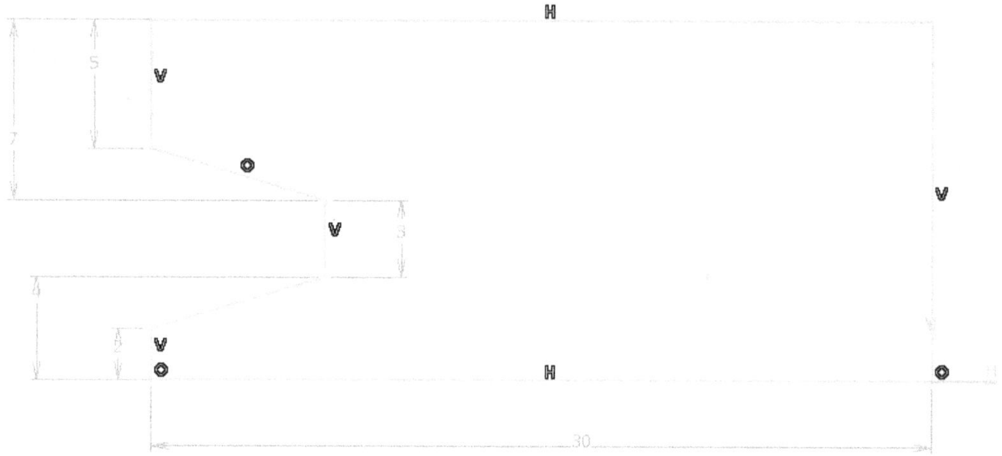

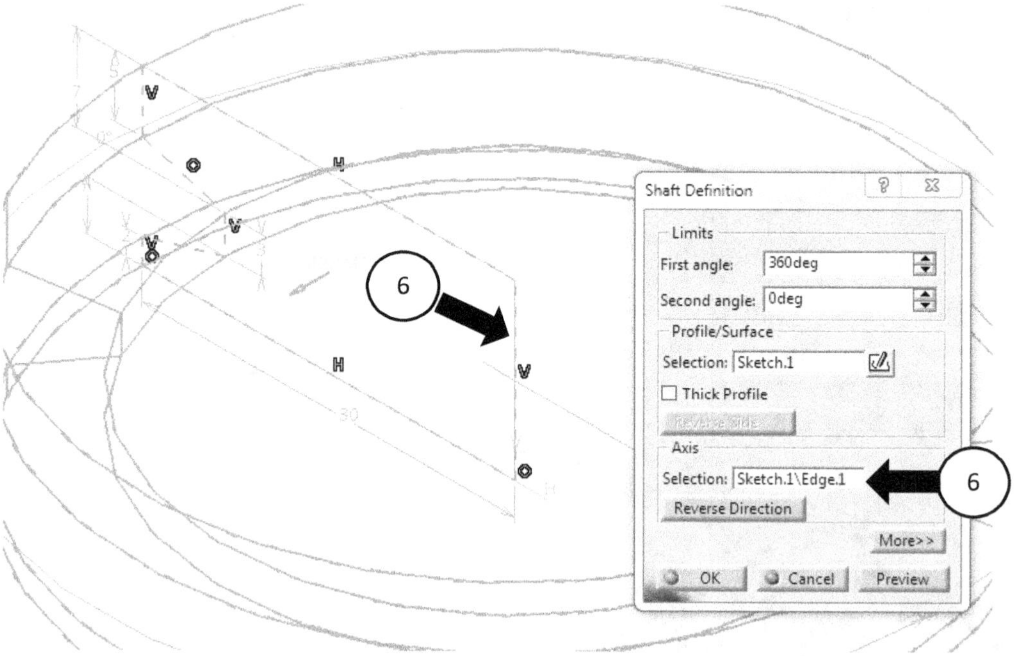

Let's practice revolving.

1. Open a new part.
2. Set it up in inches.
3. Create a sketch as shown on the left on the YZ plane—ensure the origin is located on the left lower corner.
4. Exit the Sketcher.
5. Click on the Shaft icon
6. Select the vertical line for the Axis.
7. Click OK.

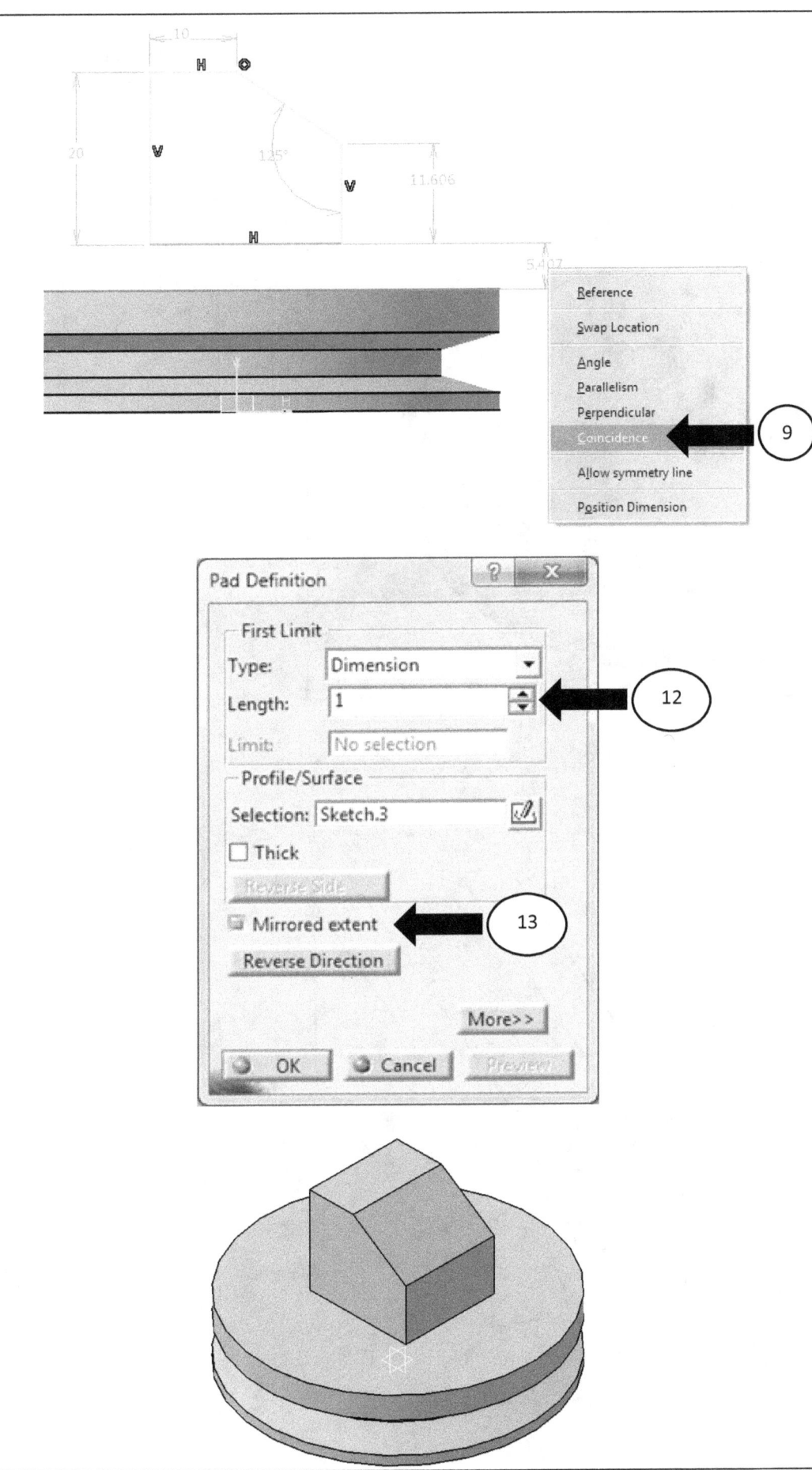

8. Create a sketch on the YZ plane – dimension shown on the left.
9. Select "Coincidence" by right-clicking when constraining the bottom line to the surface as shown on the left.
10. Exit the Sketcher.
11. Click the Pad icon.
12. Enter 1 in Length.
13. Activate "Mirrored extent".
14. Click OK.

It should appear similar to the sketch on the left.

35

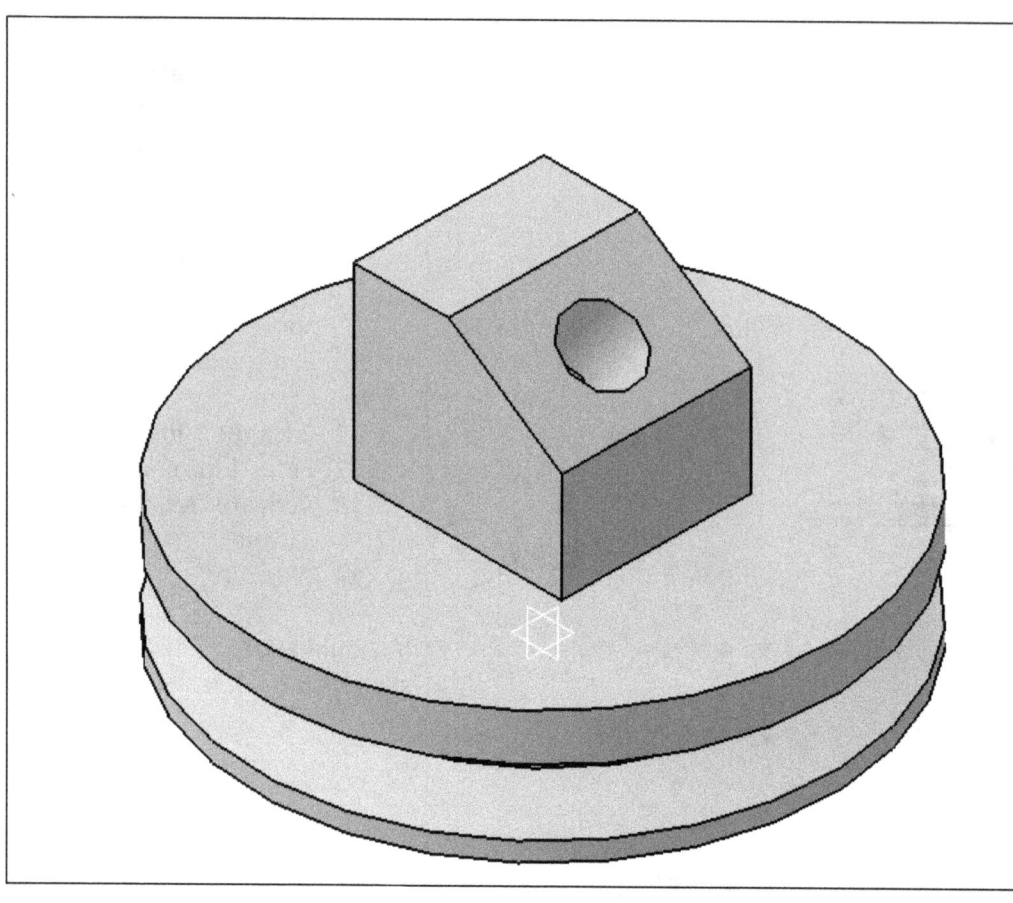

Let's add a hole on the sloped surface. Try to locate the hole in the center. Diameter is .4, the depth is .5.

The result should be similar to the sketch on the left.

Save as "Revolve."

Chapter 3 Assignment

1: Unit is mm

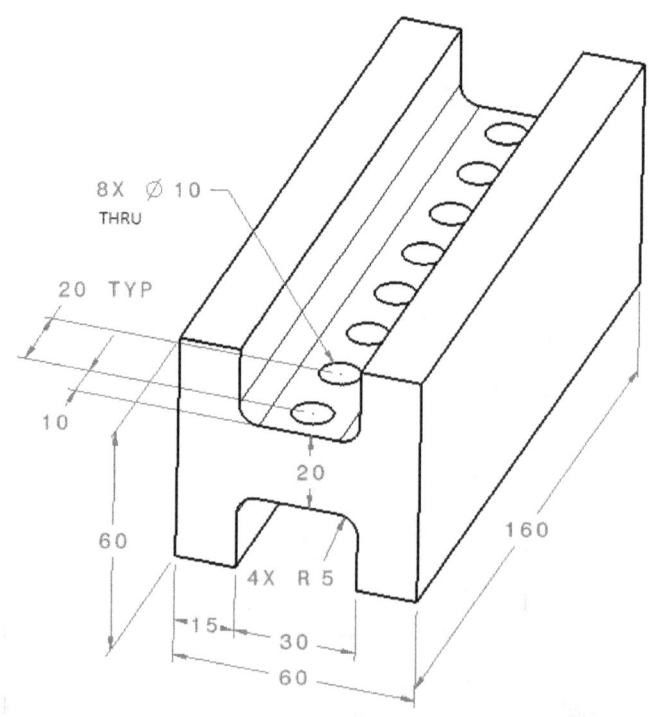

36

2: Unit is inches

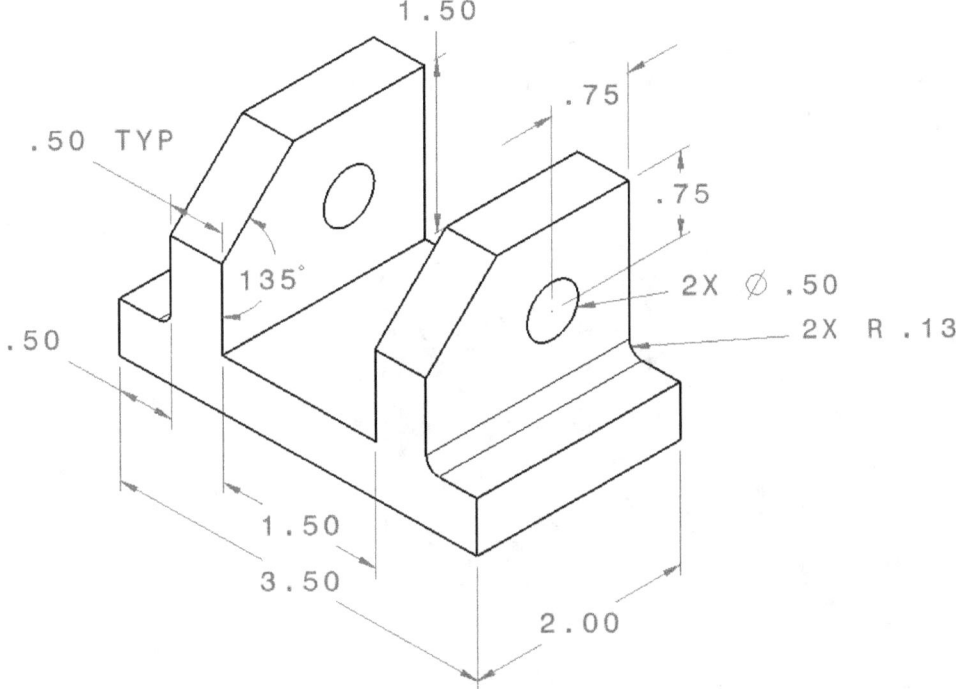

3: Unit is inches. Size and scale are up to you

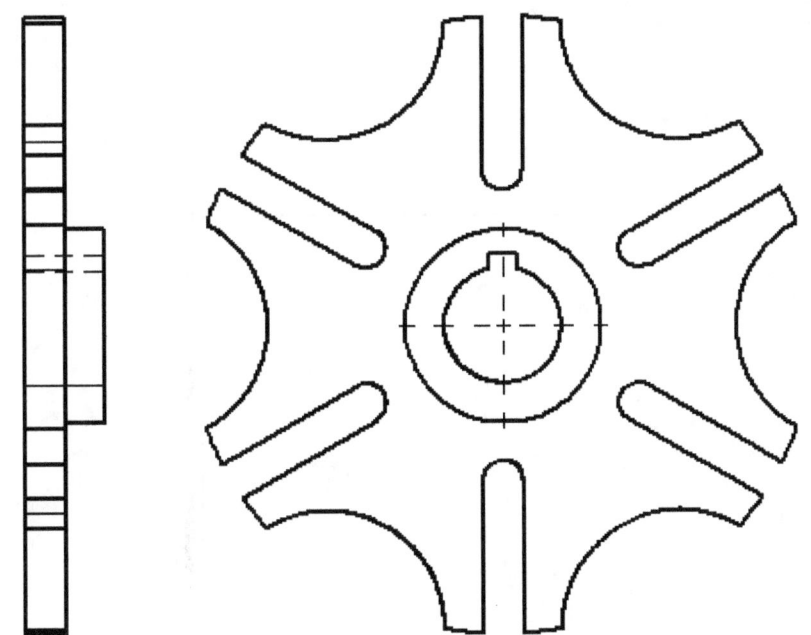

Chapter 4 – Drafting Workbench

The default software set-up is not correct in order to do drafting in Catia V5 at school. You will receive two Drafting files in inches and mm from the instructor. Please save them on your PC or in a thumb drive. They will also be available in the Class drive in ET labs.

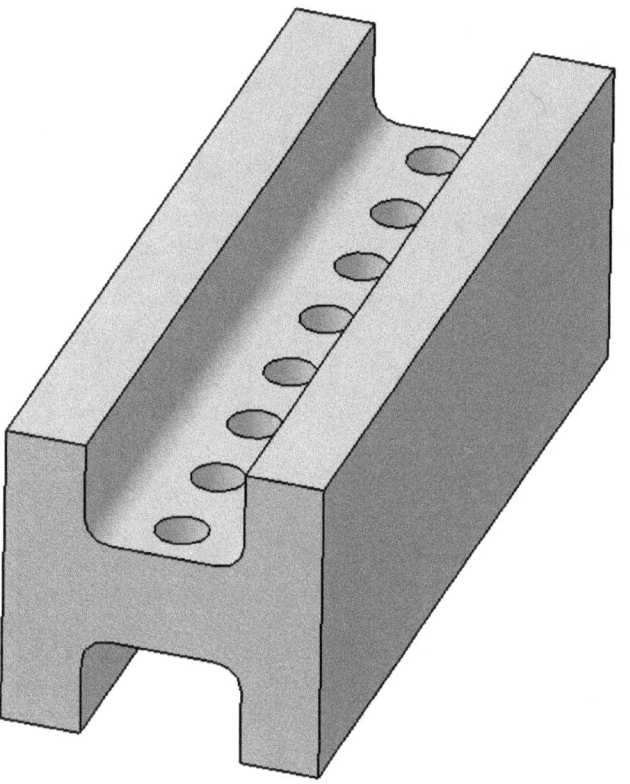

Open Chapter 3 Assignment 1. Also open the Drafting mm file.

In order to create a drawing, the Views toolbar is very important. Under the Front View icon, there are more functions to create certain views.

1. Click on the Front View icon 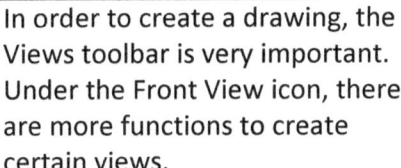.
2. Go to the Part.
3. Click on one of the surfaces – this brings you back to the drafting page.

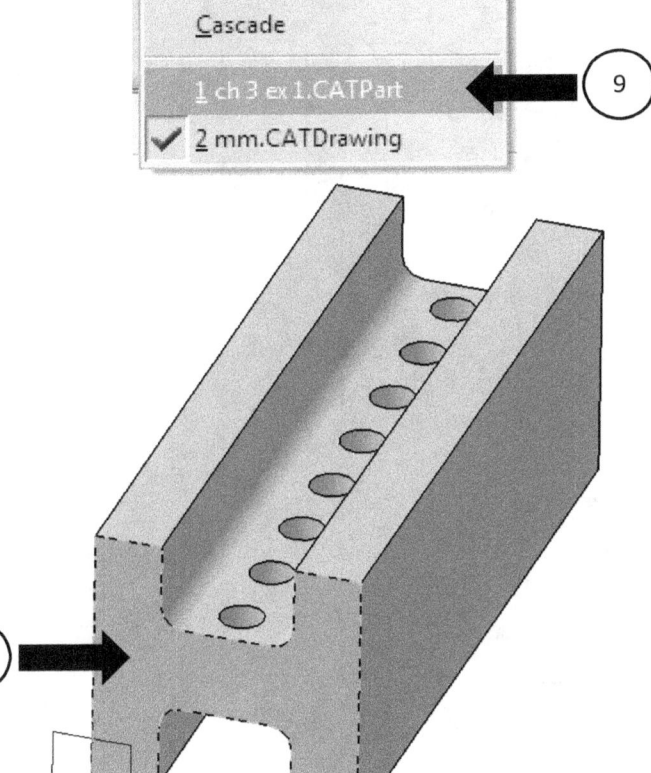

On the right upper corner, you will see a blue compass. This switches the view orientations.

If you like the view, click anywhere.

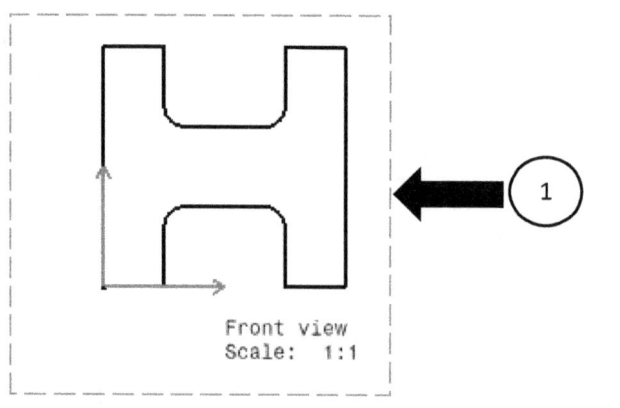

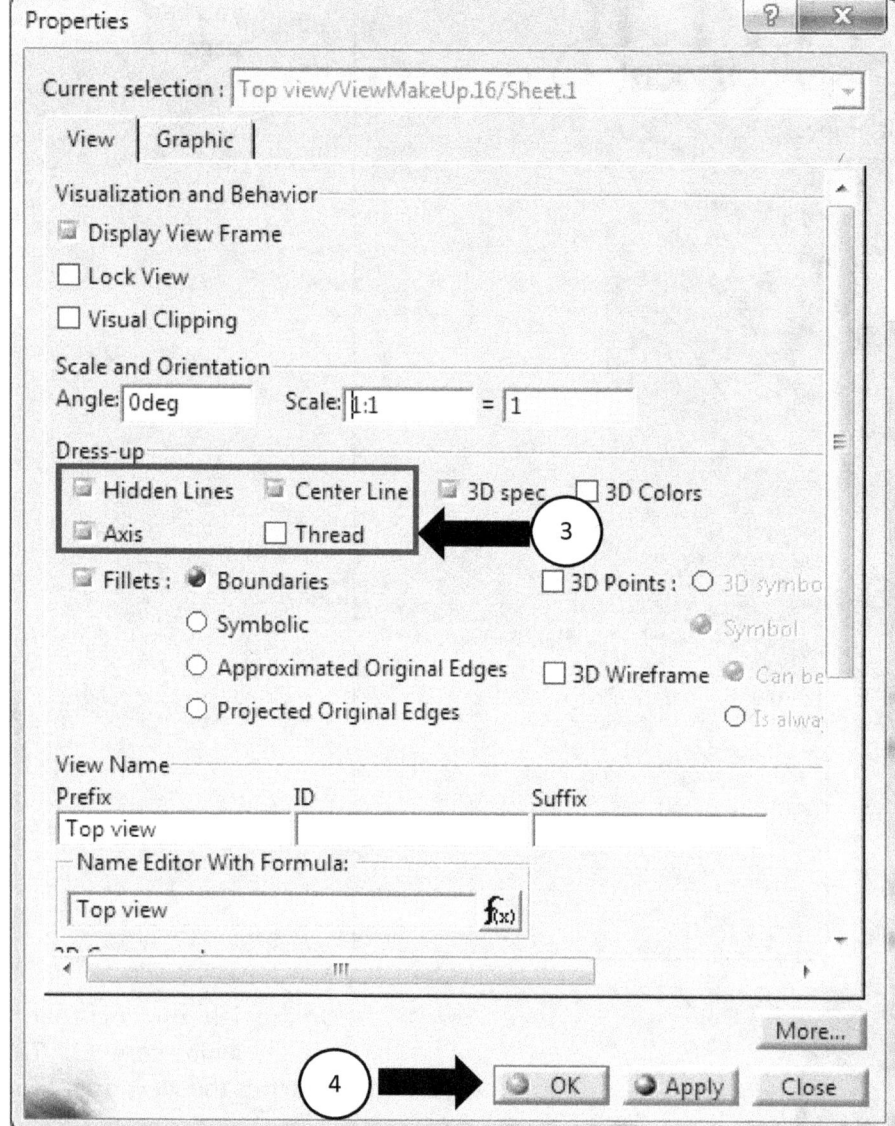

As shown on the left, a view will be created.
This view needs hidden lines and center/axis lines.

1. Right click on the dotted line frame around the view.
2. Click Properties.
3. Activate Hidden lines, Center line, and Axis boxes.
4. Click OK.

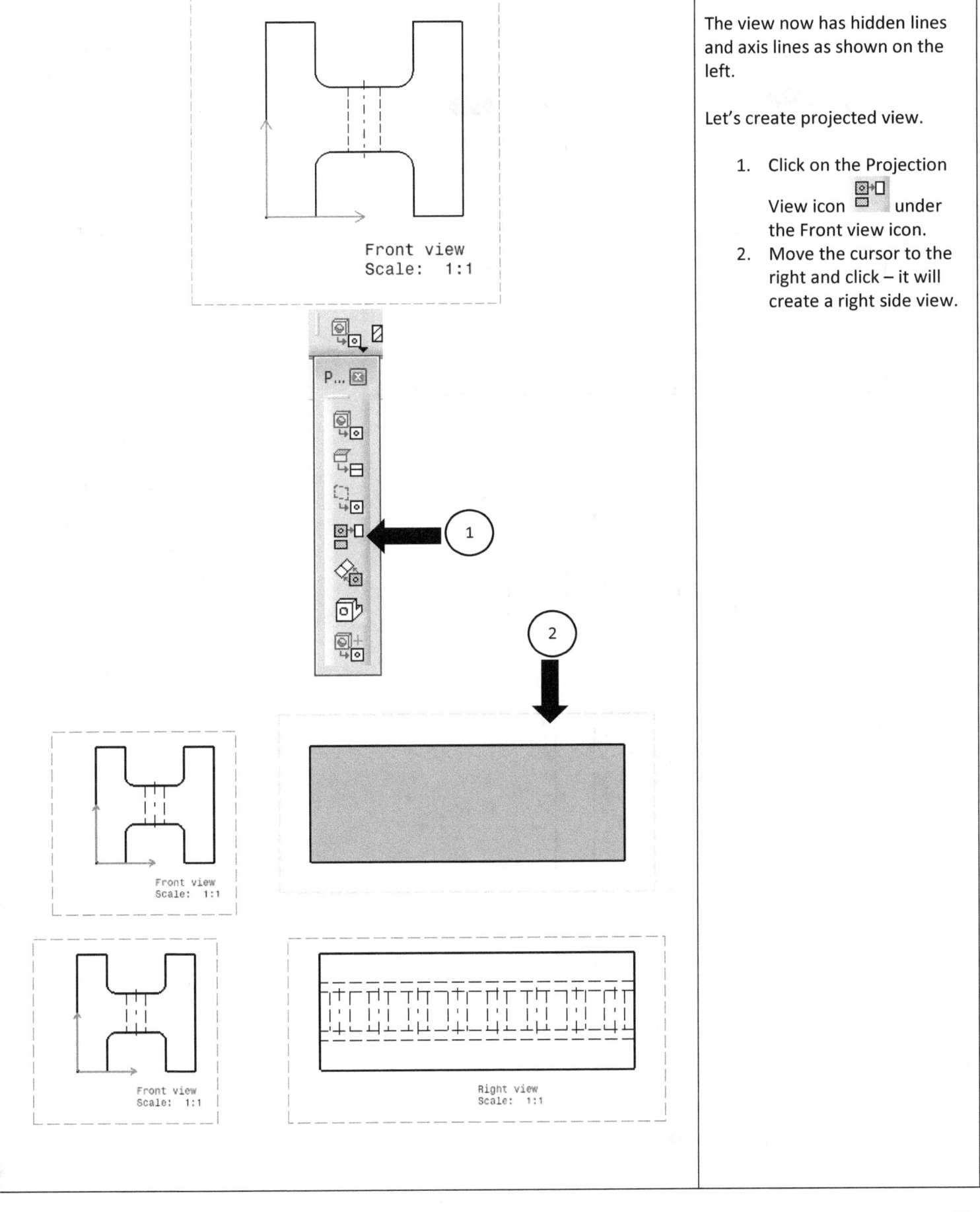

The view now has hidden lines and axis lines as shown on the left.

Let's create projected view.

1. Click on the Projection View icon under the Front view icon.
2. Move the cursor to the right and click – it will create a right side view.

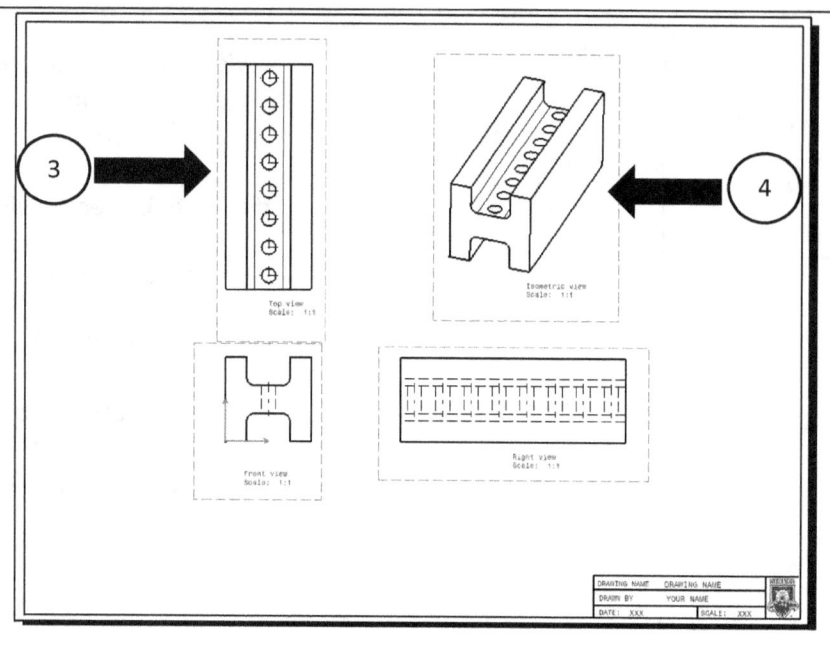

3. Repeat the same step and create a top view.
4. Click on the Isometric View icon and create an isometric view.

Your drawing should be similar to the sketch on the left.

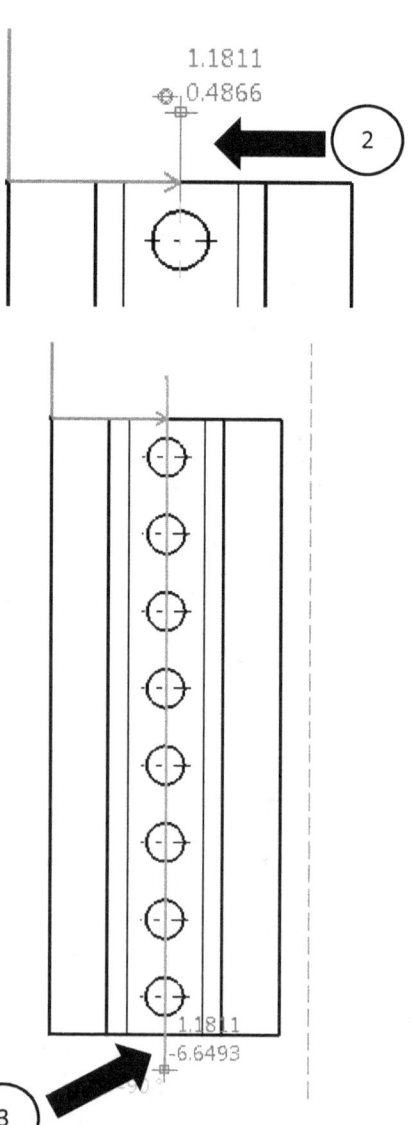

Let's add a section view.

1. Click on the Offset Section View icon.
2. Move the cursor to the center of the part – this creates a guiding line as shown on the left.
3. Click and drag a line all the way through.

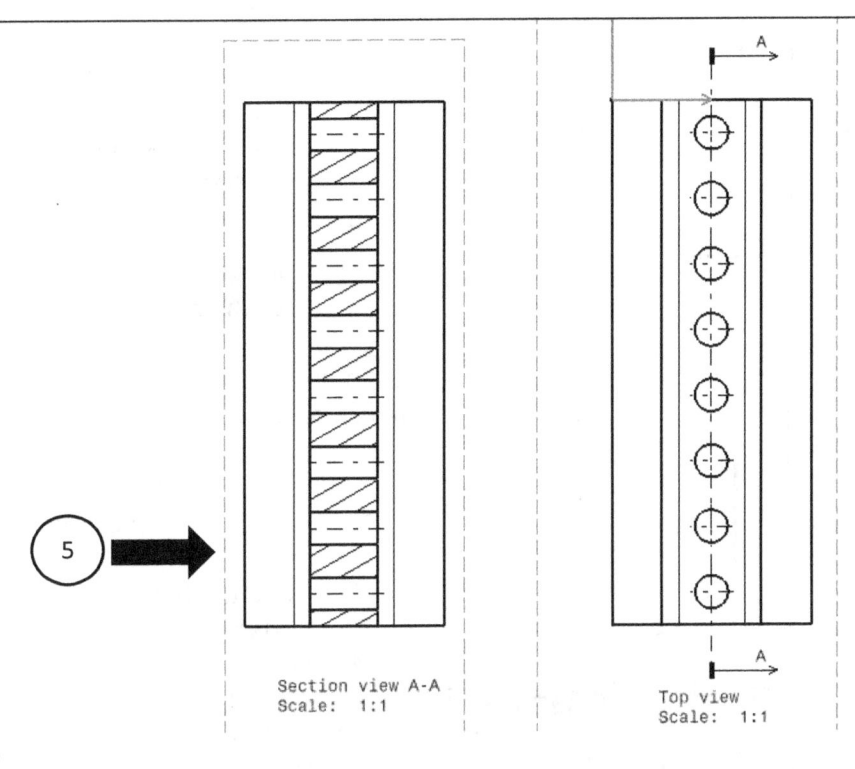

4. Double-click at the end.
5. Move the cursor to the left and click.

It should look similar to Section view A-A.

Let's add dimensions. The Dimensions toolbar as shown on the left is needed.

To add dimensions, make sure to activate the view by double-clicking the frame. The dotted line frame should be red – this means the view is activated.

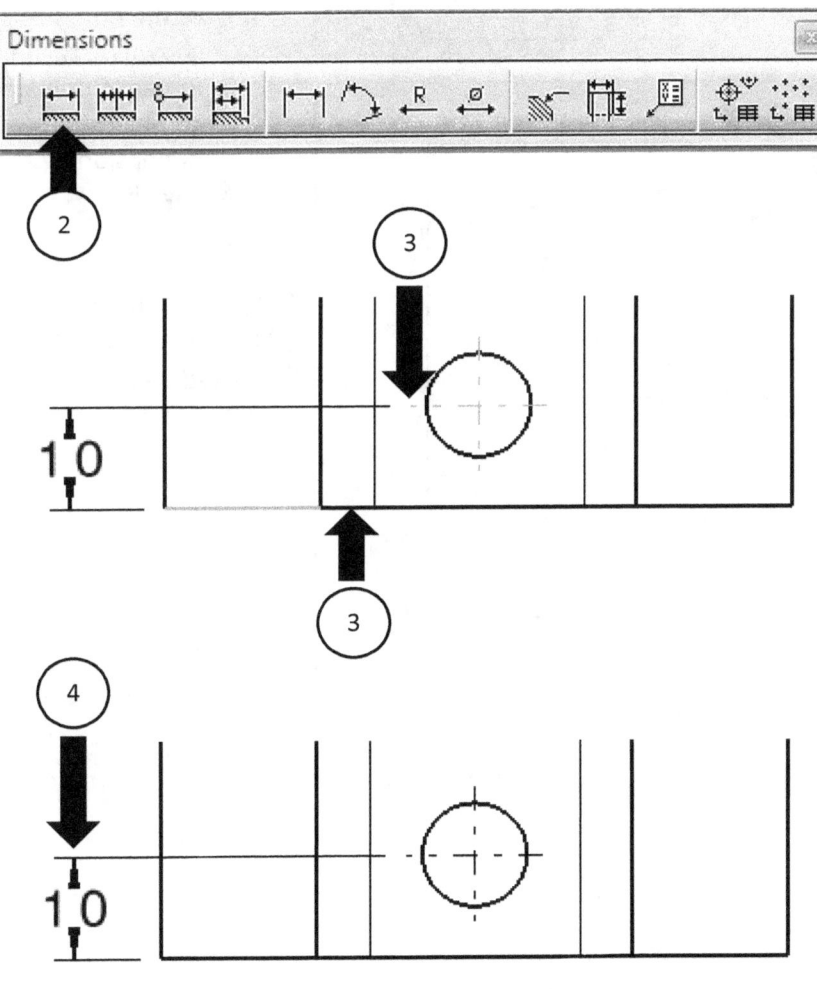

1. Activate the top view by double-clicking the frame.
2. Click on the Dimensions icon.
3. Click the bottom line and the center line – 10mm should appear as shown on the left.
4. Click where you want to place it.

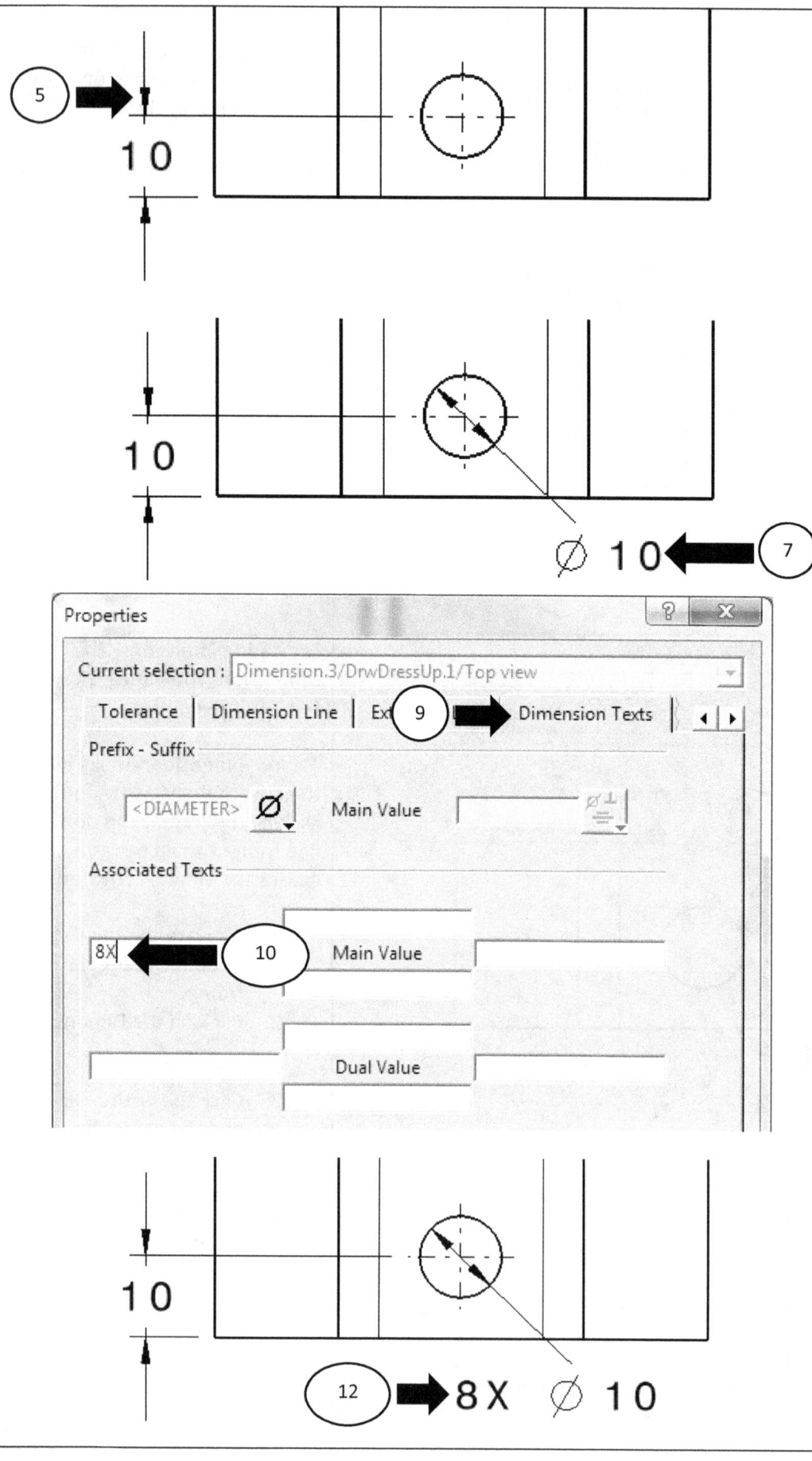

5. The arrowheads need to be switched to the outside – click the arrowhead
6. Click on the diameter icon.
7. Click the circle.

It should now be similar to the sketch on the left

8. Right click on the hole size dimension.
9. Go to the Dimension Texts tab.
10. Add "8X" in front of the Main Value.
11. Click OK.
12. "8X" was added in front of the hole size as shown on the left.
13. Repeat these steps, complete the dimensioning.
14. Save it.

Remember the dimensioning rules such as: no over-dimensioning, no dimensioning on hidden lines...

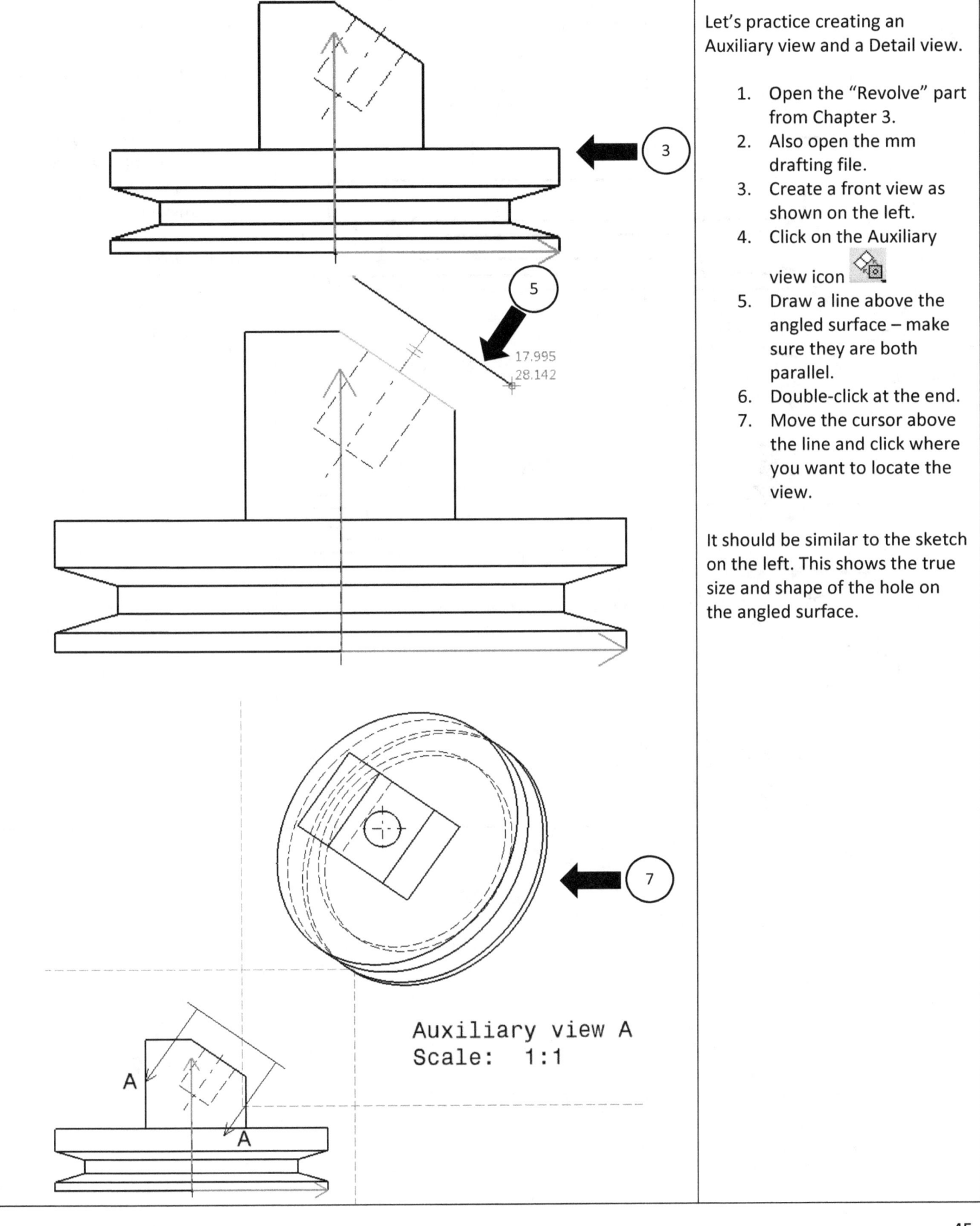

Let's practice creating an Auxiliary view and a Detail view.

1. Open the "Revolve" part from Chapter 3.
2. Also open the mm drafting file.
3. Create a front view as shown on the left.
4. Click on the Auxiliary view icon
5. Draw a line above the angled surface – make sure they are both parallel.
6. Double-click at the end.
7. Move the cursor above the line and click where you want to locate the view.

It should be similar to the sketch on the left. This shows the true size and shape of the hole on the angled surface.

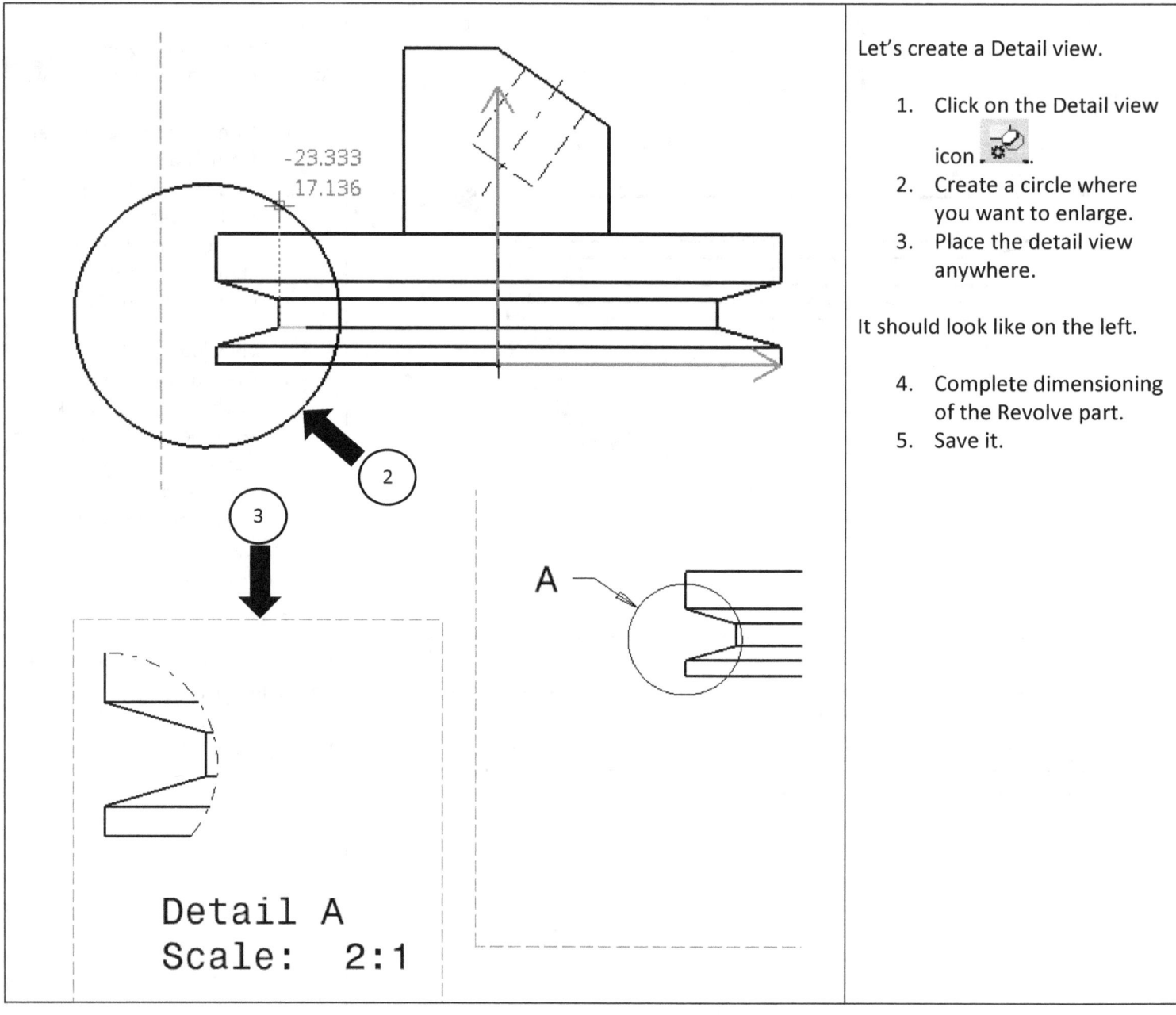

Let's create a Detail view.

1. Click on the Detail view icon.
2. Create a circle where you want to enlarge.
3. Place the detail view anywhere.

It should look like on the left.

4. Complete dimensioning of the Revolve part.
5. Save it.

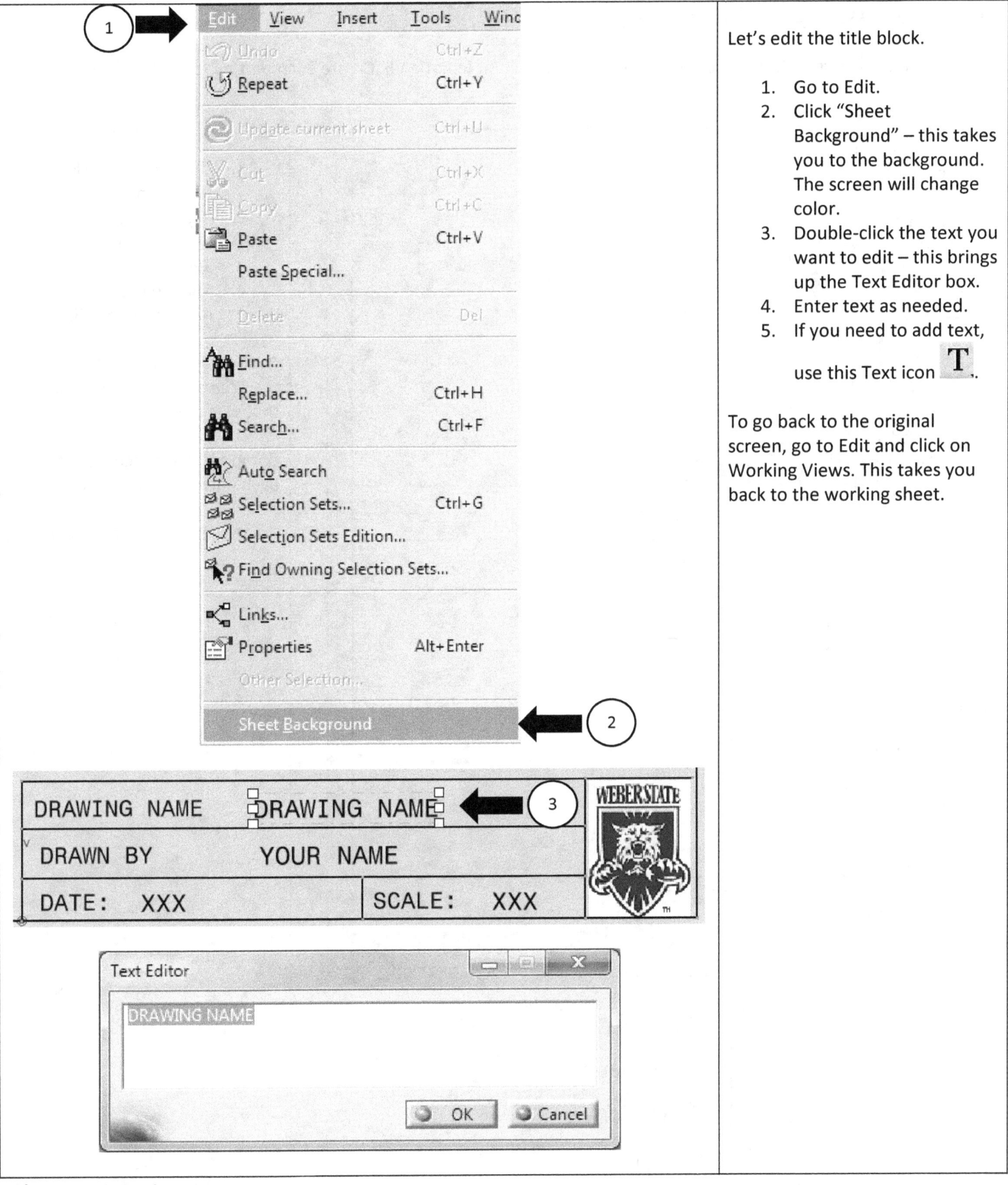

Let's edit the title block.

1. Go to Edit.
2. Click "Sheet Background" – this takes you to the background. The screen will change color.
3. Double-click the text you want to edit – this brings up the Text Editor box.
4. Enter text as needed.
5. If you need to add text, use this Text icon T.

To go back to the original screen, go to Edit and click on Working Views. This takes you back to the working sheet.

Chapter 4 Assignment

1. Create a drawing of Chapter 3 assignment 2.

47

Chapter 5 – Assembly Design Workbench

When creating an assembly, naming each part is very important in Catia V5. Create each part by following the step below.

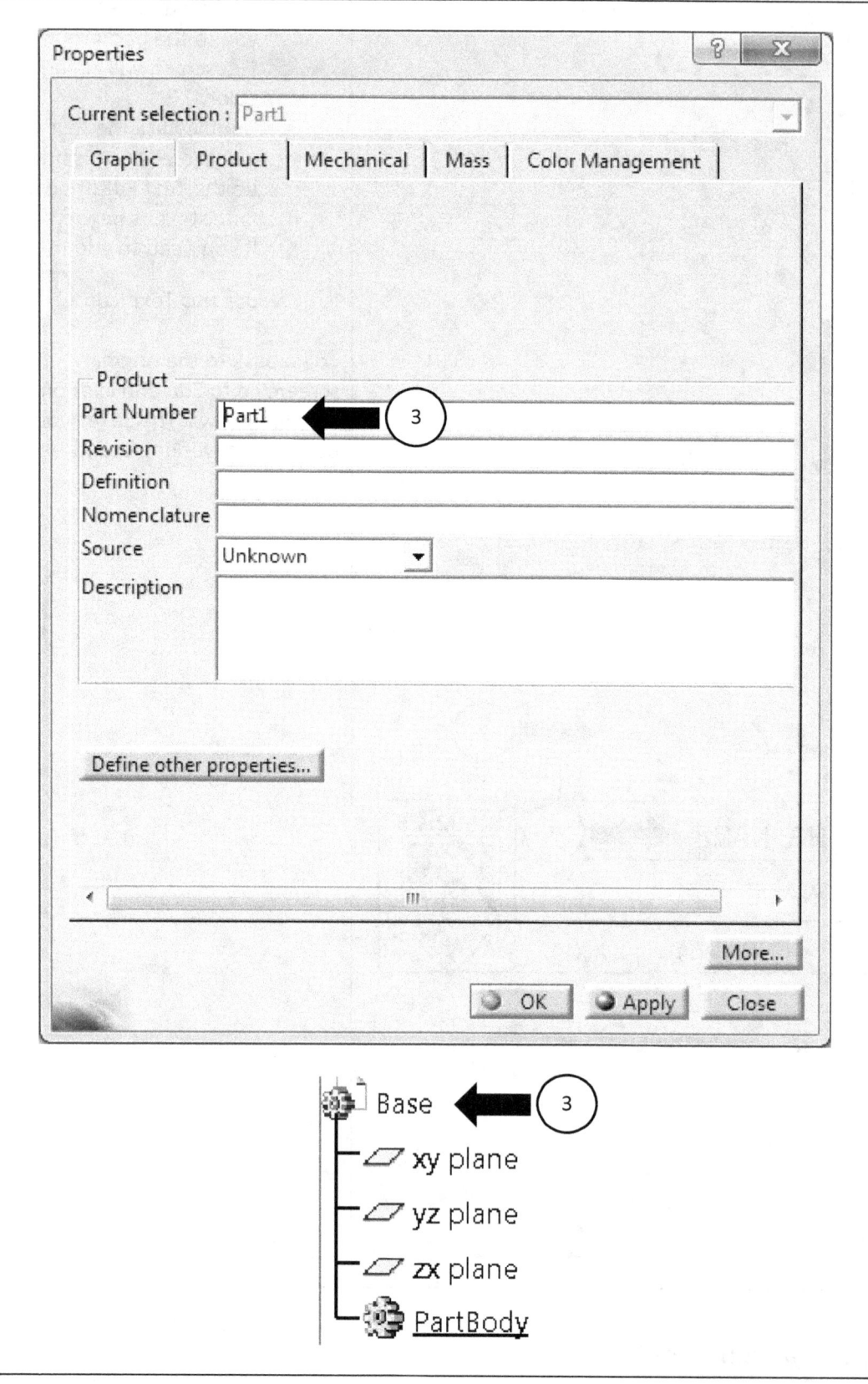

1. Open a Part Design Workbench
2. Right click on the Part 1 on the Tree.
3. Select Properties.
4. Change the Part Number name to a specific name.

As shown on the left, the top of the Tree says the specific name that was typed.

Complete this procedure for each part in this chapter.

Create each part shown below.

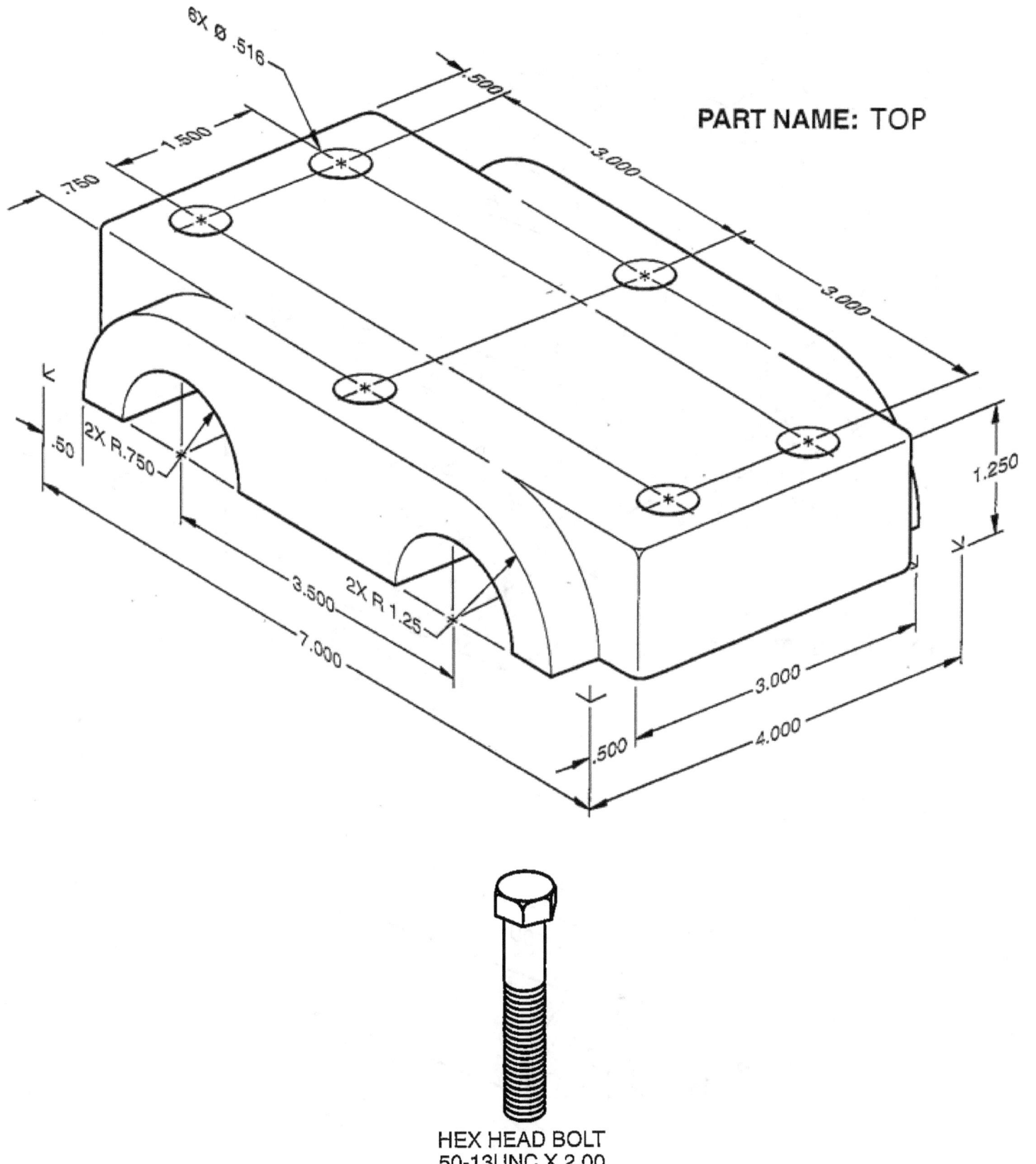

PART NAME: TOP

HEX HEAD BOLT
.50-13UNC X 2.00

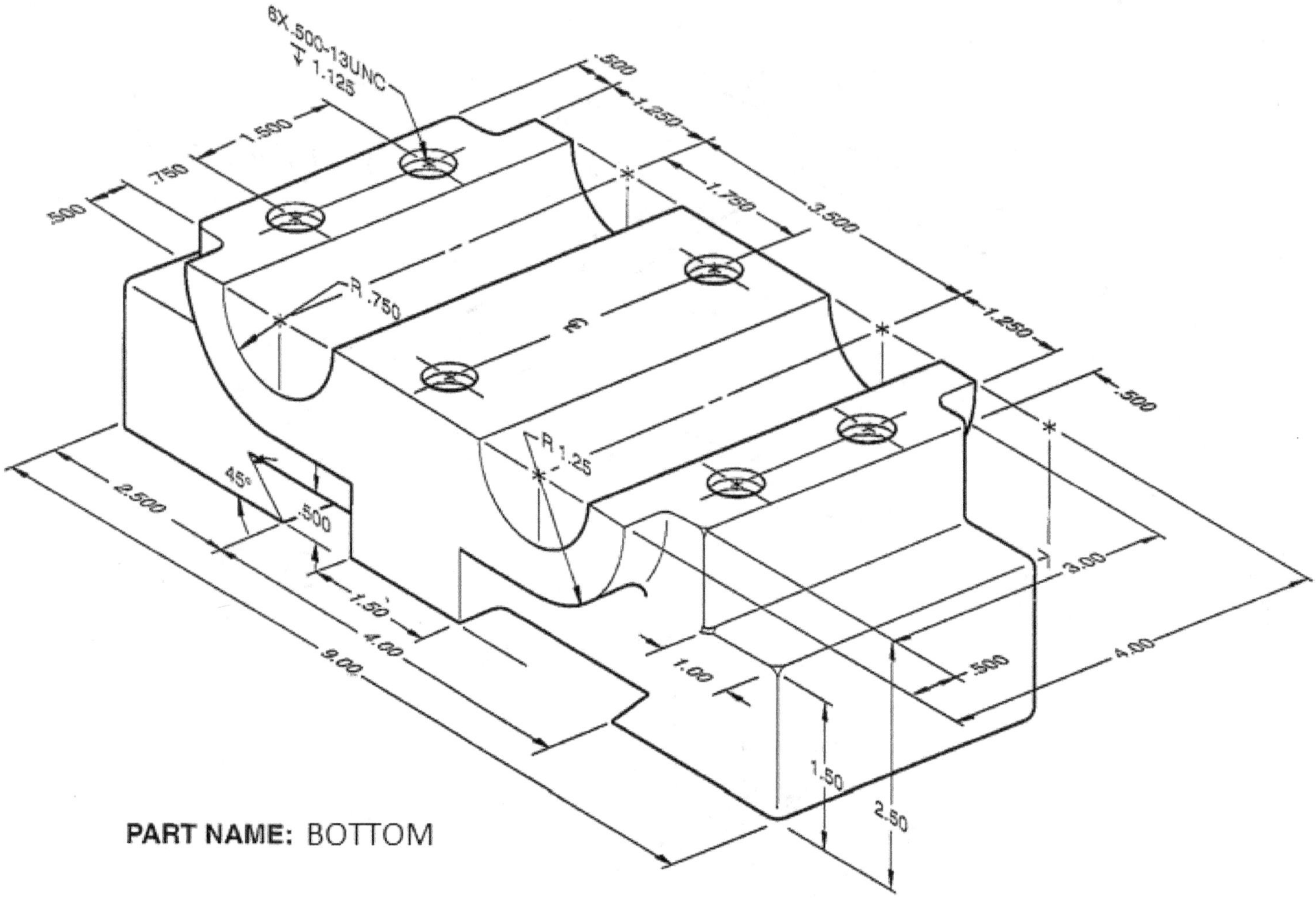

PART NAME: BOTTOM

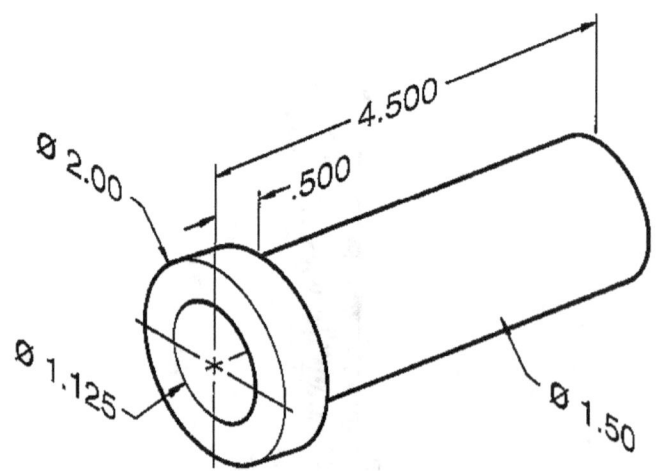

PART NAME: BUSHING

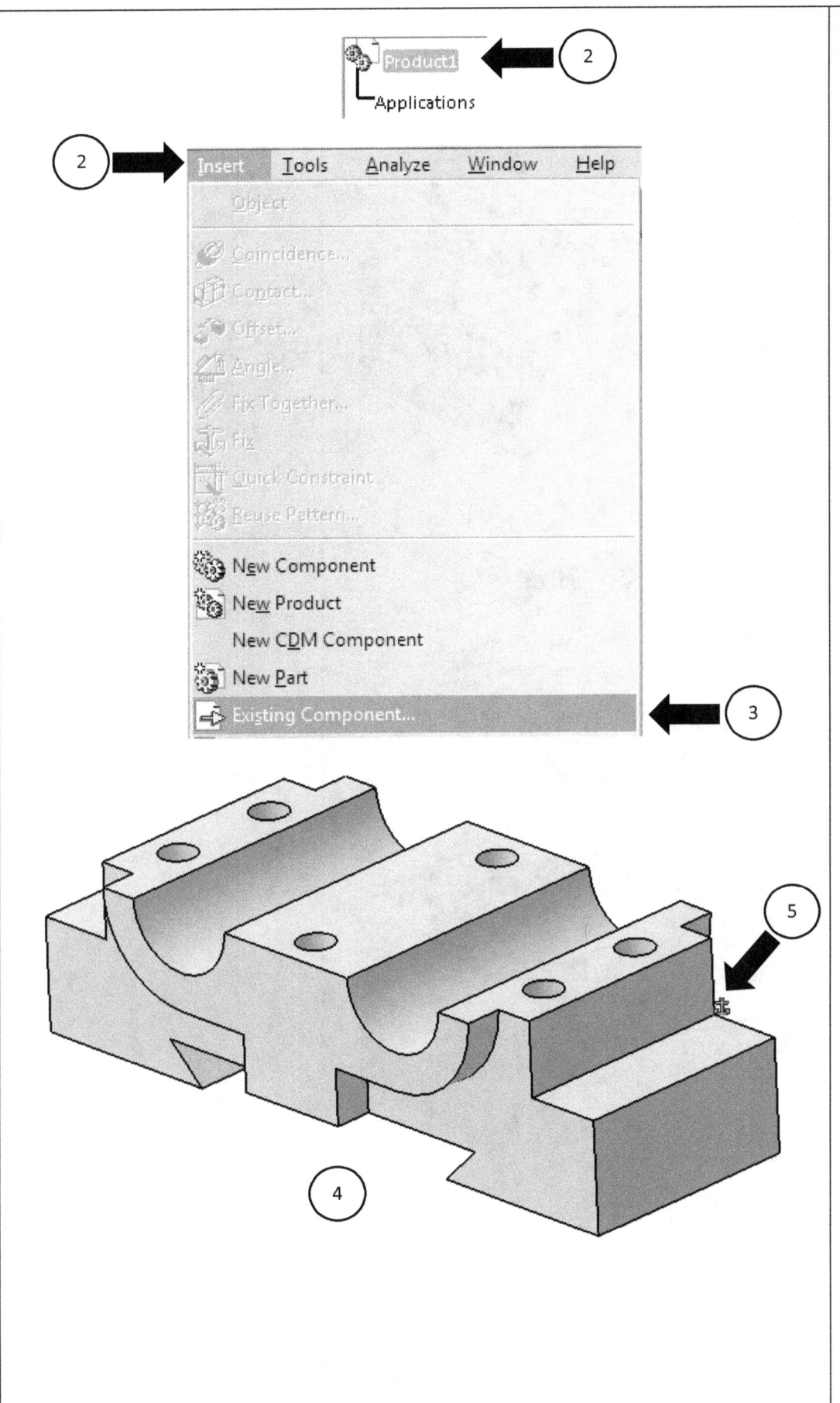

Now let's put each piece together.

1. Go to Start, Mechanical Design, Assembly Design.
2. Click on the "Product 1" on the Tree to highlight it.
3. Go to Insert, Existing Component, and select the Bottom that you just created.
4. Rotate around the part with the Compass to place the Bottom correctly as shown on the left.
5. Click on the Fix icon and click on the Bottom. The Bottom is now fixed and will stay as is.

51

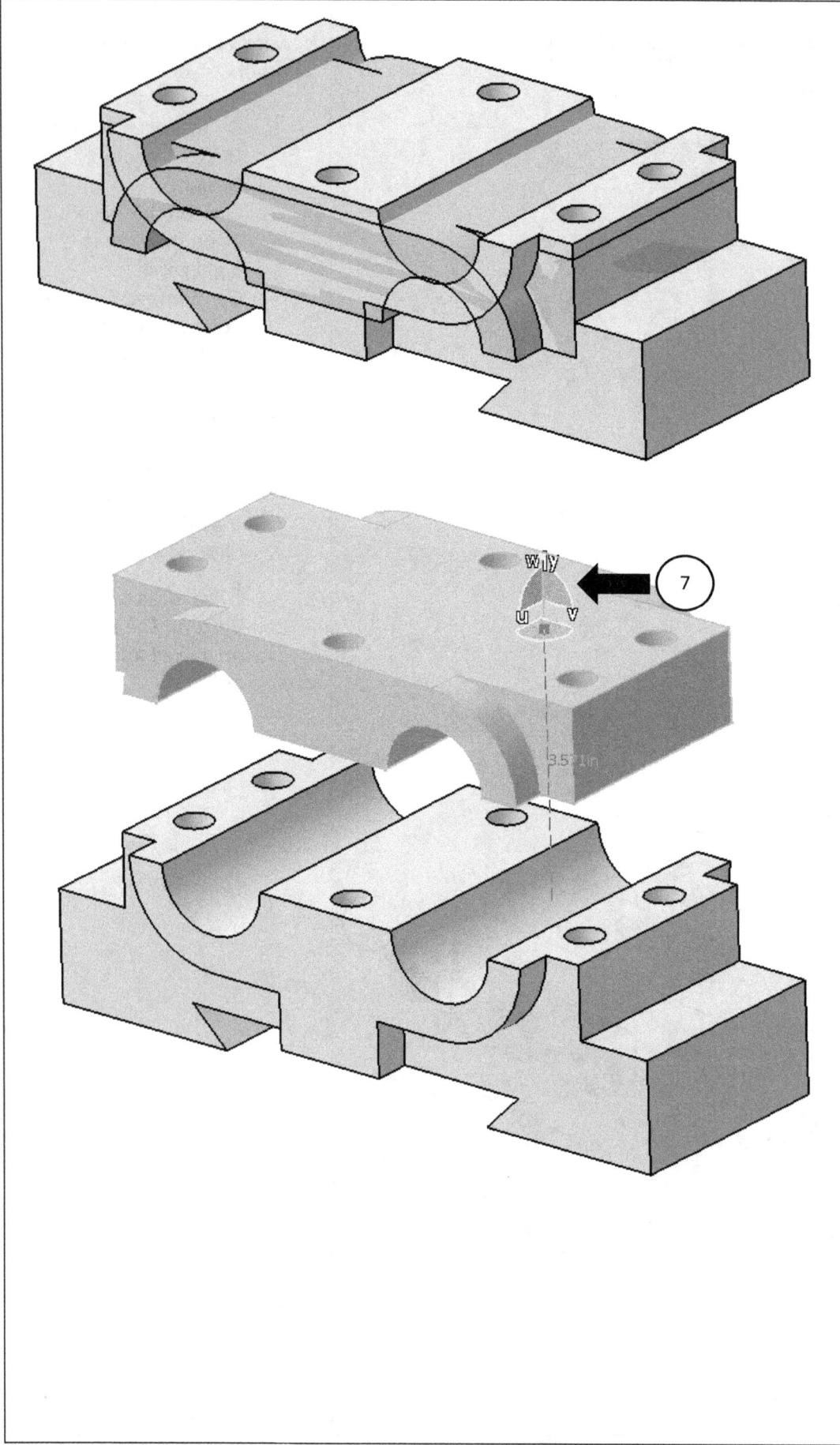

6. Repeat the same step, insert the Top.
7. If the Top comes in being overlapped with the Bottom, use the Compass to pull it from the Bottom as shown on the left.

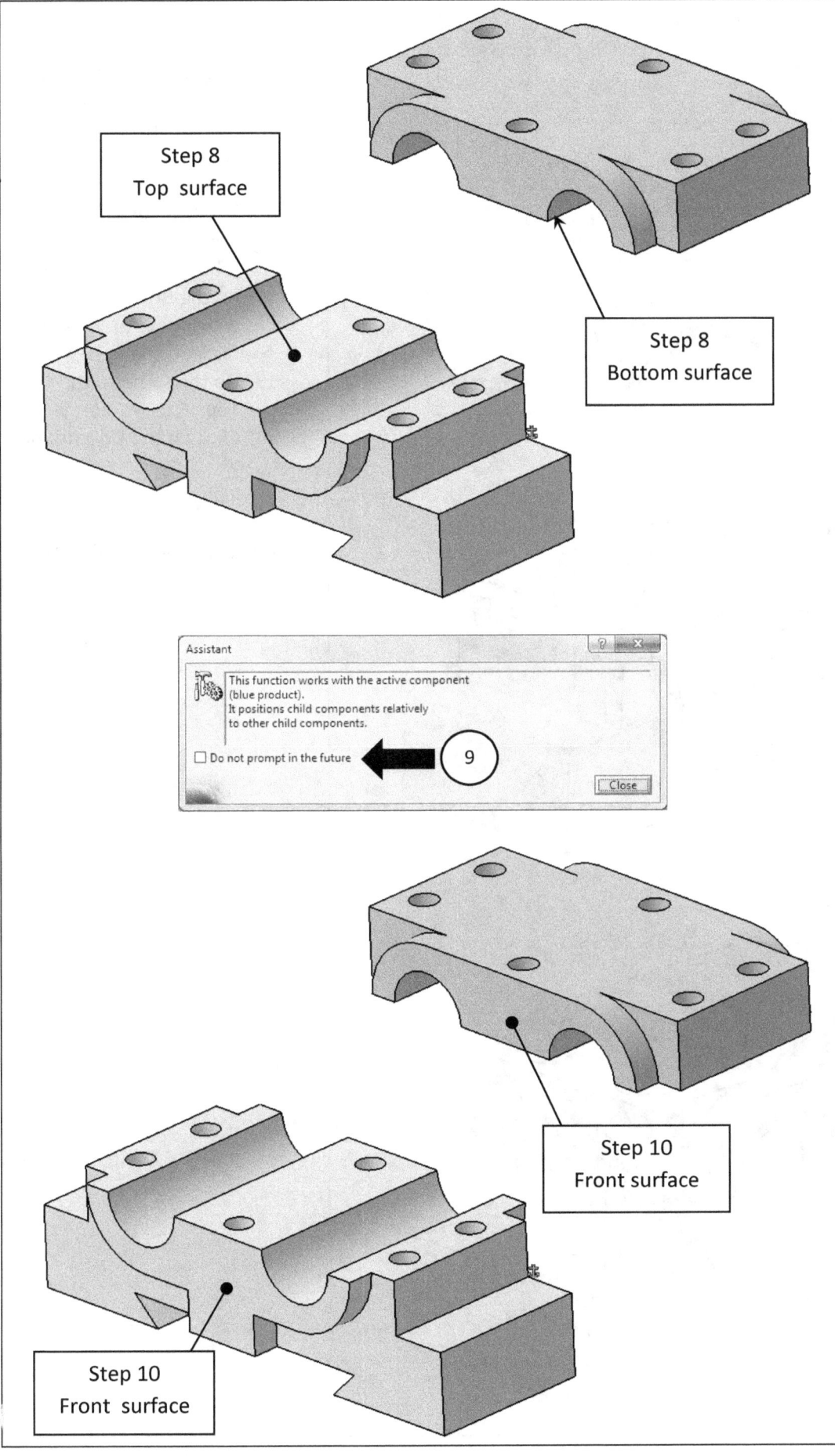

8. Use the Contact Constraint icon and click on the surface as shown on the left.

9. If the Assistant window shows up, click on the "Do not prompt in the future" and click Close.

10. Click on the Coincident Constraint icon and click on the front surfaces as shown on the left.

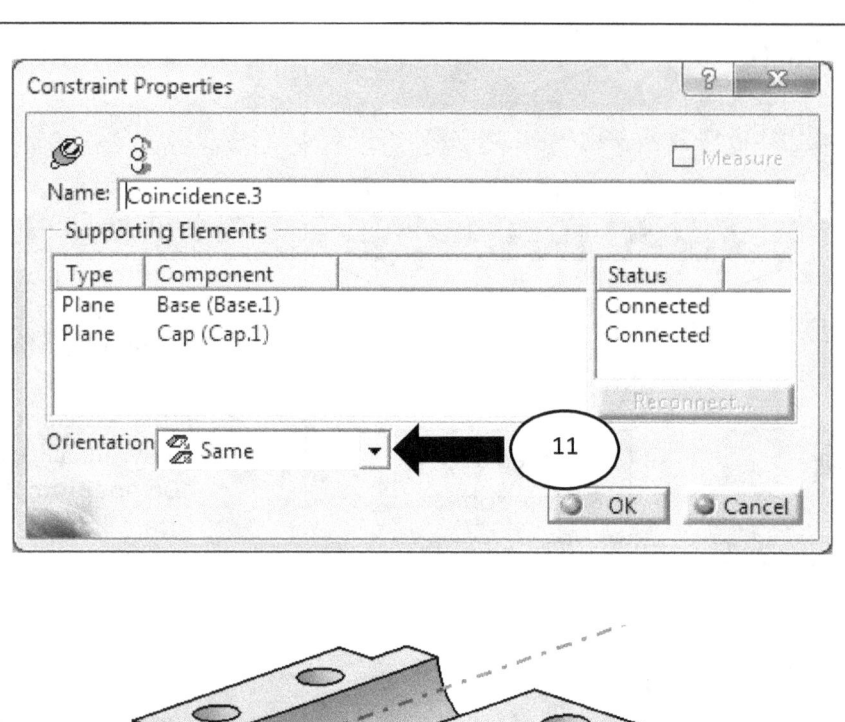

11. When the Constraint Properties box shows up, ensure the Orientation says "Same." Click OK.
12. Use the Coincidence Constraint icon to have it hover over half hole on the Bottom – this will appear an Axis line. Click on the Axis.
13. Do the same to the Top.
14. Click on the Update All icon .

The result should look like on the left.

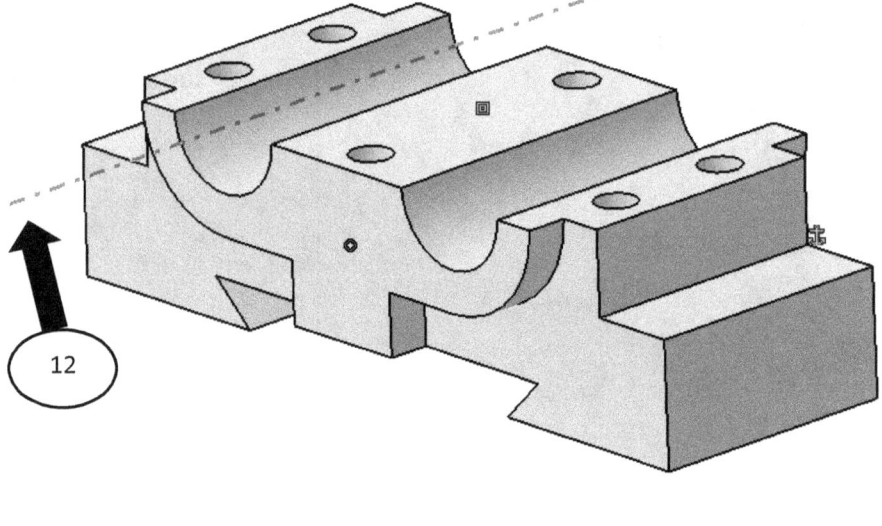

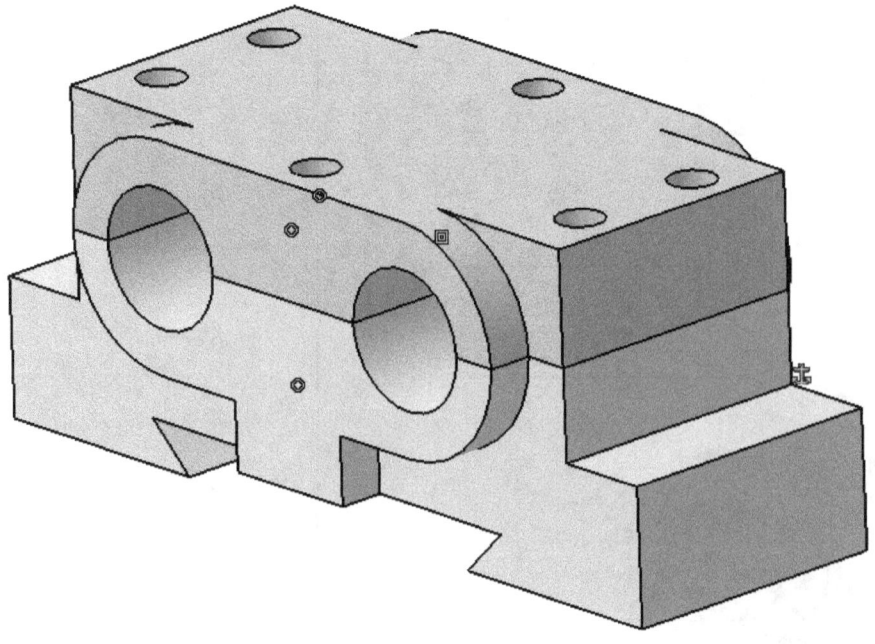

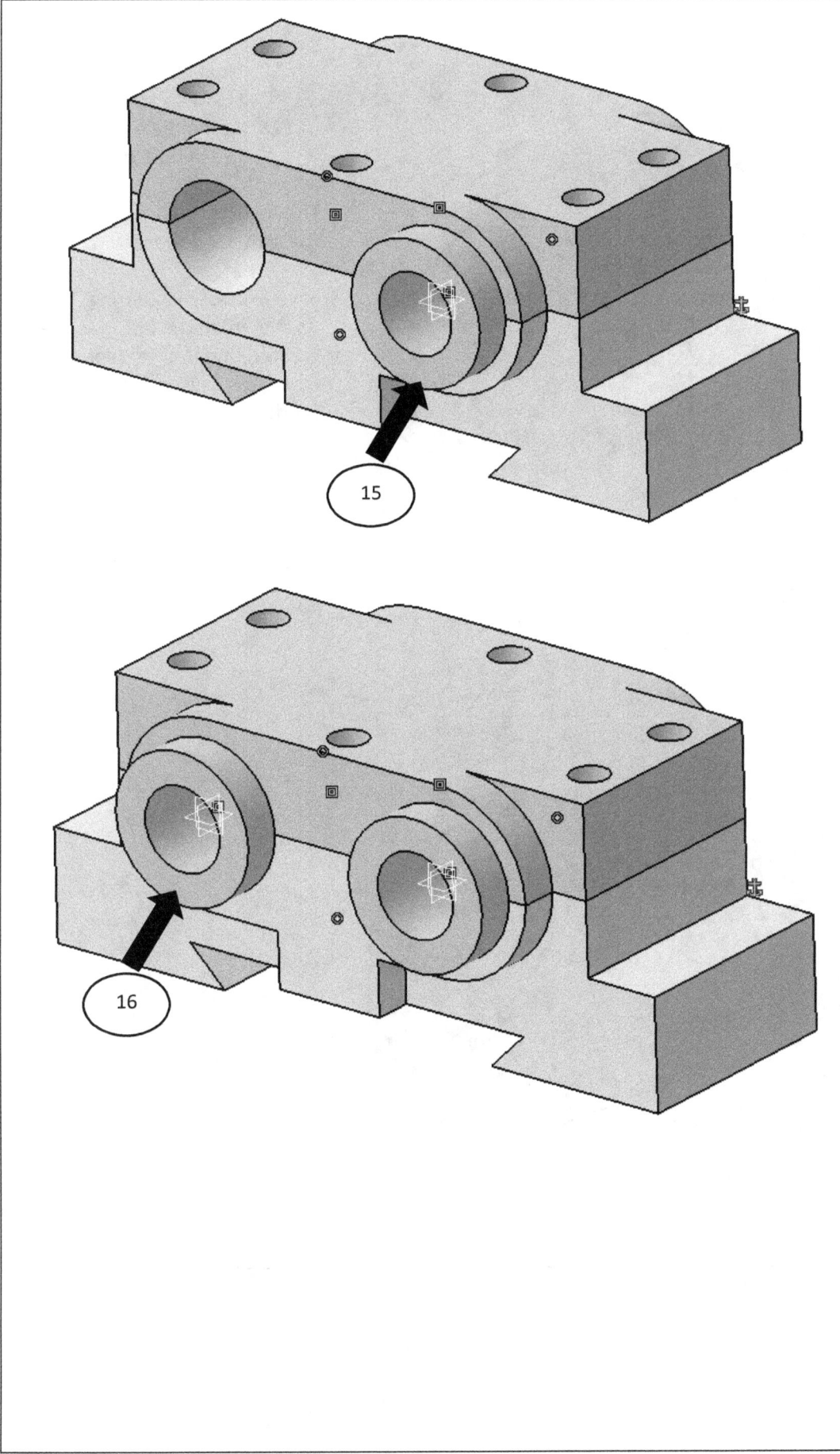

15. Insert the Bushing and use the previous steps and assemble it as shown on the left.
16. Repeat the Step 13 again and insert one more Bushing and assemble it as shown on the left.

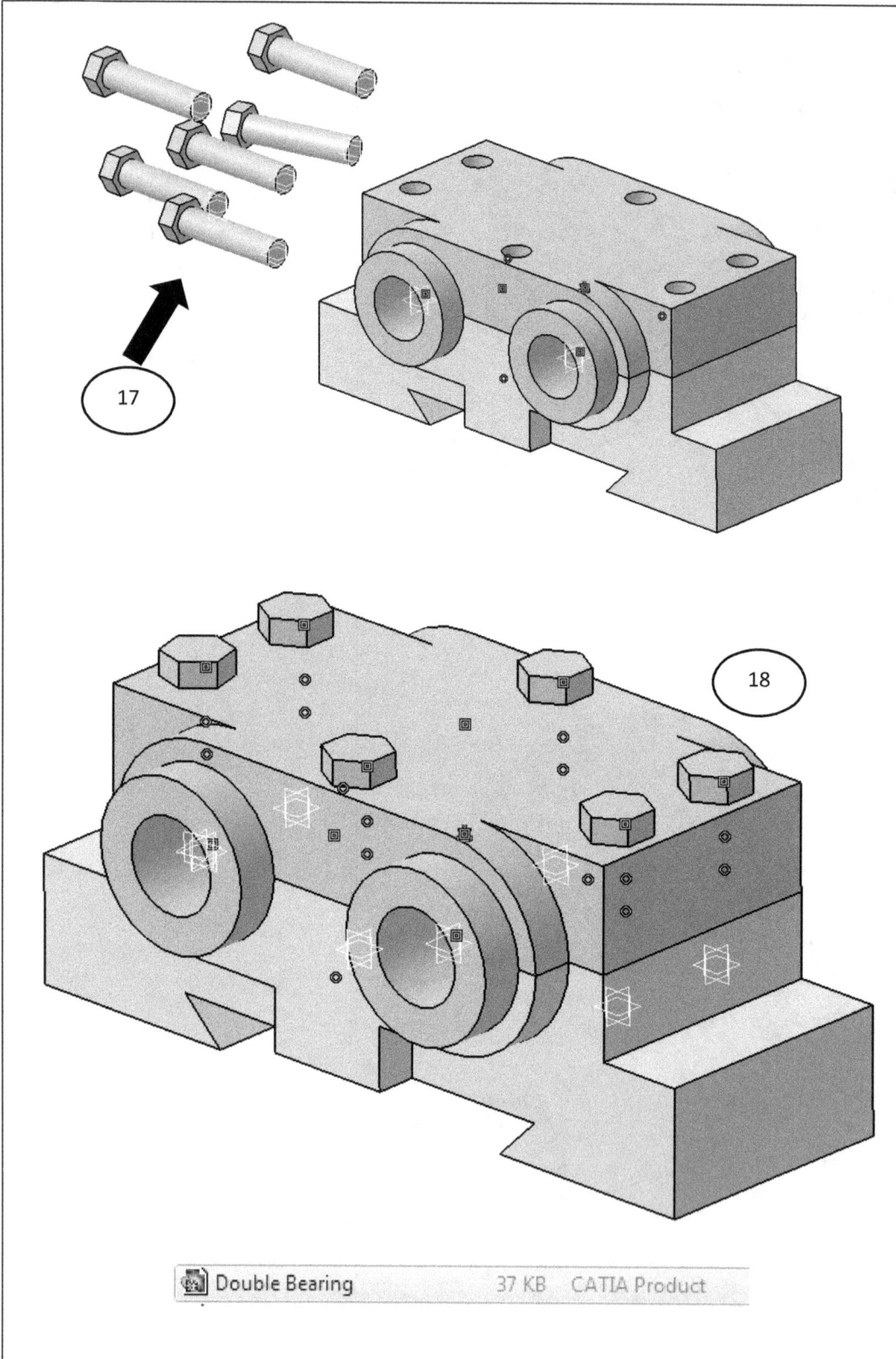

17. Next, insert the Hex Head Bolt six times.
18. Assemble each bolt into a hole on the Top.

The result should be similar to the sketch on the left.

Save as "Double Bearing"
As shown on the left, this was created under a Product 1; the extension says "CATIA Product."

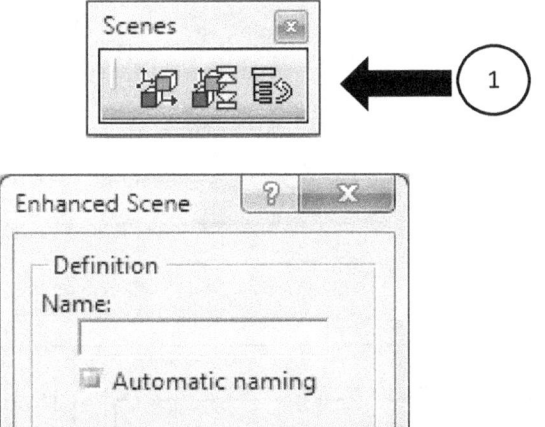

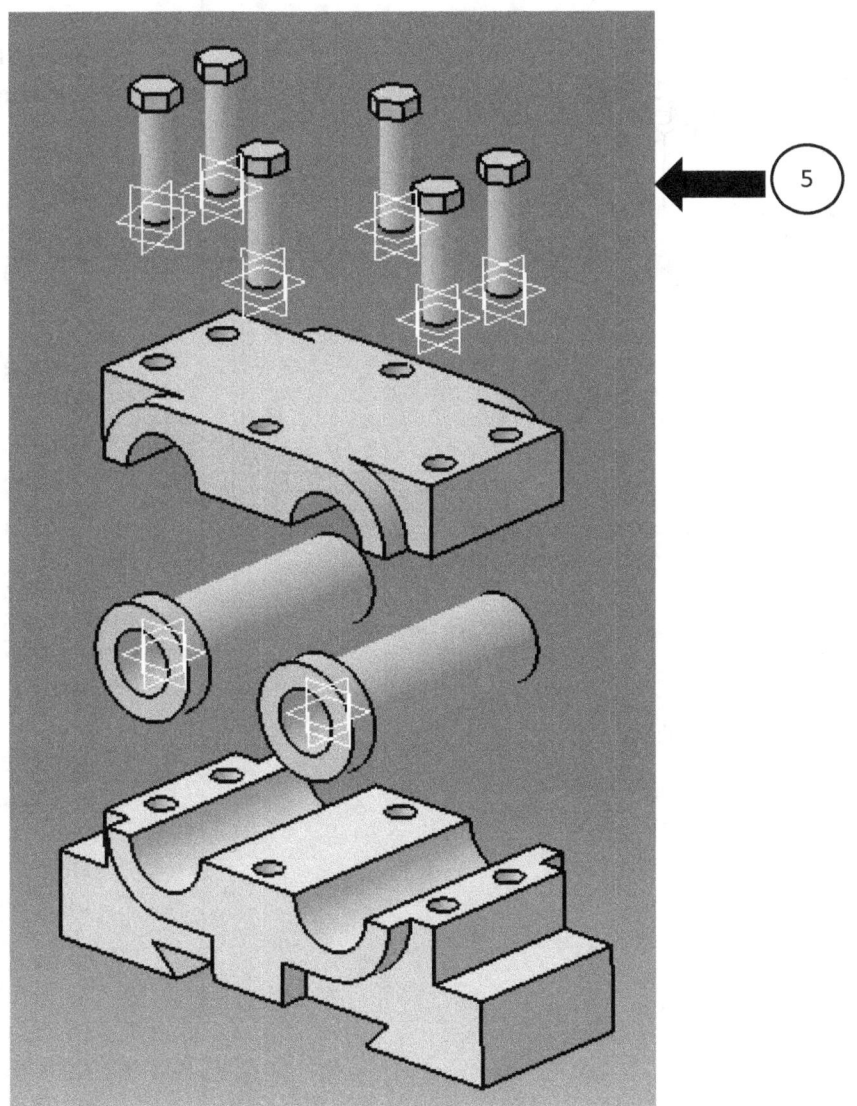

Let's create an exploded view of this assembly.

1. Open the Scenes toolbar.
2. Click on the Enhanced Scene icon .
3. Select "Full" on the box.
4. Click OK.

This brings you to a different space – the background color has changed.

5. In this space, by using the Compass, pull each part from each other as shown on the left.
6. After it is done, click on the Exit Scene icon .

57

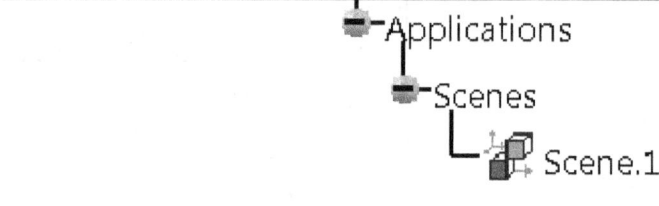

	Scene 1 is added on the bottom of the Tree as shown on the left.

Creating an assembly drawing

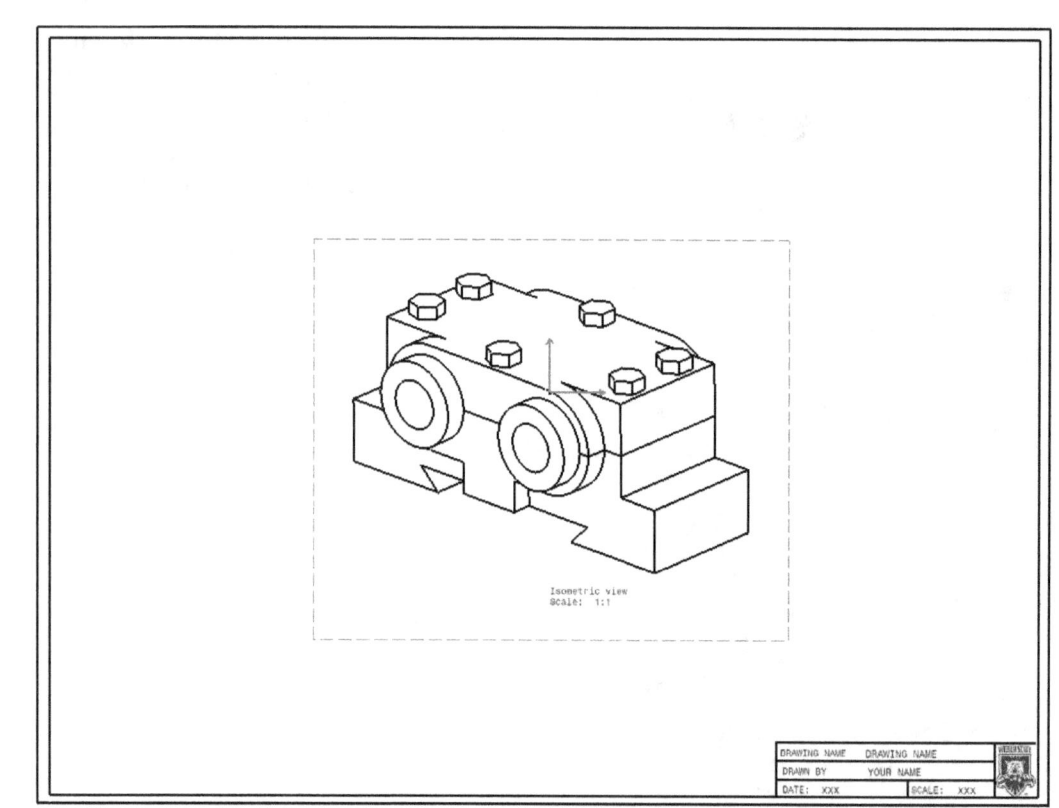

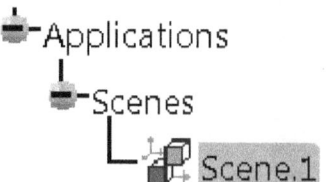

Now, let's create an assembly drawing.

1. Open the mm drawing sheet that was sent to you.
2. Make sure the assembly is also opened.
3. Click on the Isometric View icon.
4. Switch the window to the assembly product and click the front face of the Base (you can click any flat surface)
5. Switch back to the drawing sheet, and click anywhere on the sheet.

An assembled isometric view is created. If unnecessary lines are shown, hide them under the Properties.

6. Now, let's add an exploded isometric view on the same sheet.
7. Click on the Isometric View icon.
8. Switch the window to the assembly product.
9. Click the "Scene. 1" on the Tree.

58

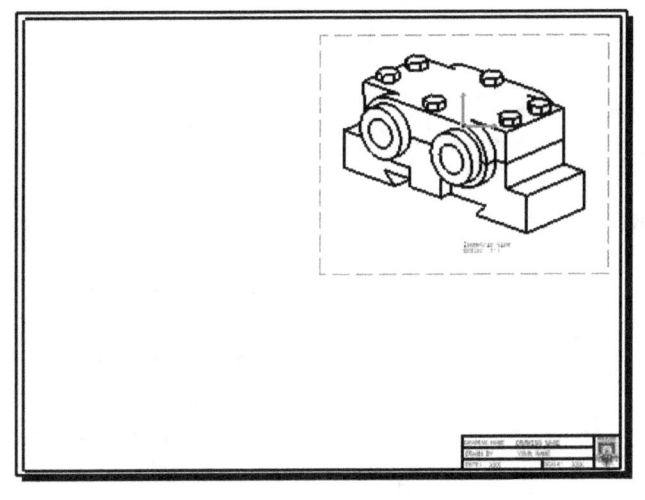
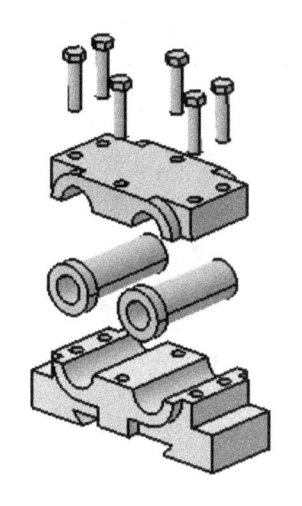

10. Click the surface of any part on the assembly.
11. That brings it back to the drawing sheet automatically, as shown on the left.
12. Click anywhere on the sheet.
13. If any unnecessary lines are shown, hide them under the Properties.

The drawing should look as shown on the left.

Now, name each view correctly.
For an assembled view, retype "Assembled Isometric View."
For an exploded view, retype "Exploded Isometric View."

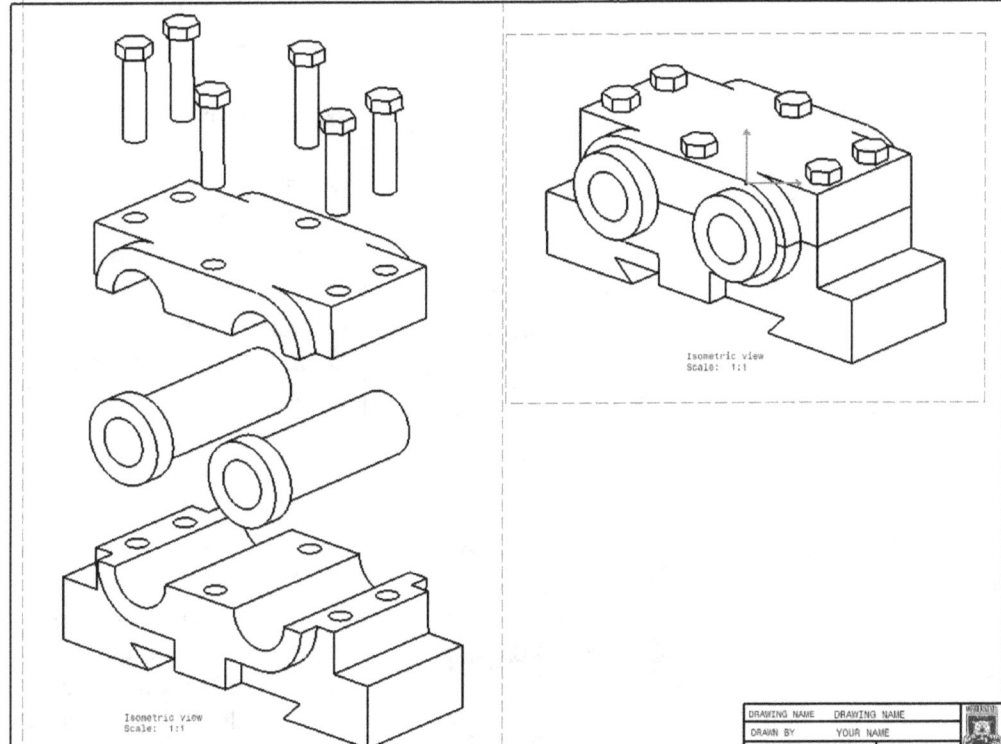

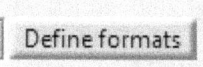

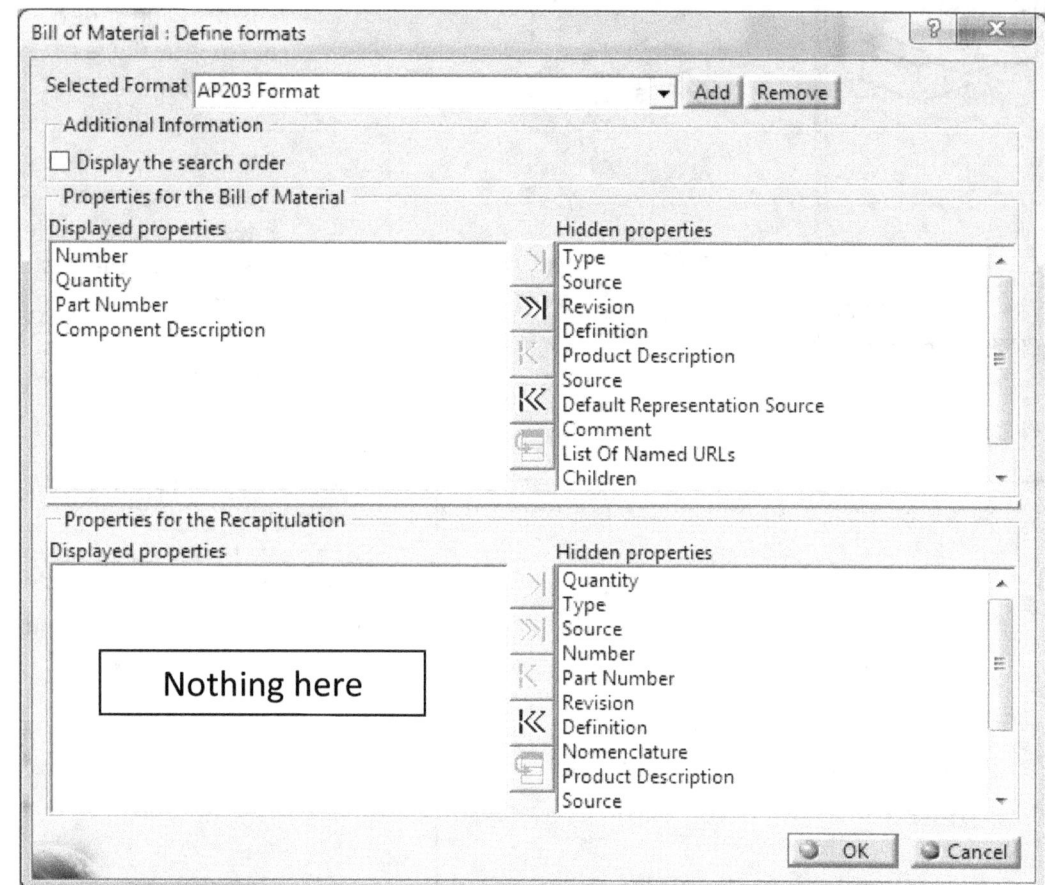

Let's insert a Bill of Material.

1. First, switch to the assembly product.
2. Go to Analyze and Bill of Material.
3. Click on the "Define formats" on the bottom right hand corner.
4. The window should look the same as shown on the left.

What you need:
Number
Quantity
Part Number
Component Description

Properties for the Recapitulation should be empty.

5. Click OK, and OK in the other window as well.
6. Go to Insert, Generations, Bill of Material, and select Bill of Material.
7. Click any empty space on the sheet.
8. BOM is inserted into the activated view frame.
9. BOM should look like on the left.

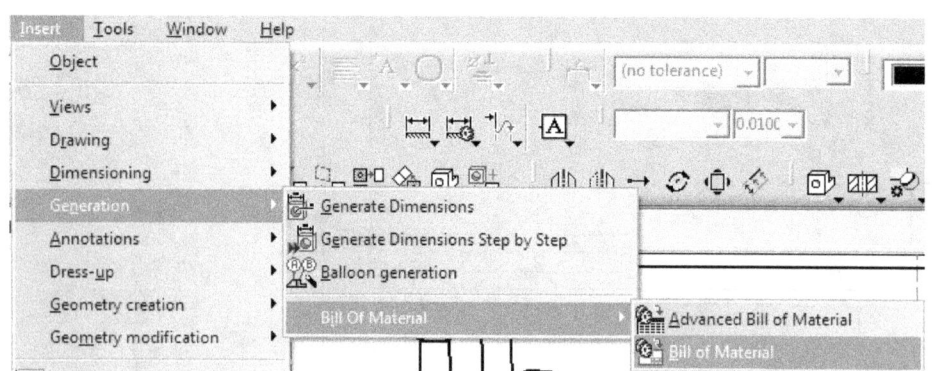

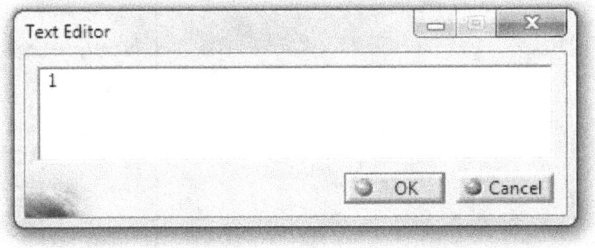

Bill of Material: Product1		
Number	Quantity	Part Num
1	1	Base
	1	Cap
	2	Bushing
	6	Screw

Bill of Material: Product1

Number	Quantity	Part Number	Component Description
1	1	Base	
2	1	Cap	
3	2	Bushing	
4	6	Hex Head Bolt	.50-13UNCX2.00

10. Double-click on the empty space across from Base, under Number and type in 1.
11. Repeat step 10 and enter 2 for Cap, 3 for Bushing and 4 for Screw.
12. Type in Screw size on the Component Description.
13. Finished BOM should look like on the left.
14. Place the BOM above the title block.

The drawing should be similar to the drawing on the left.

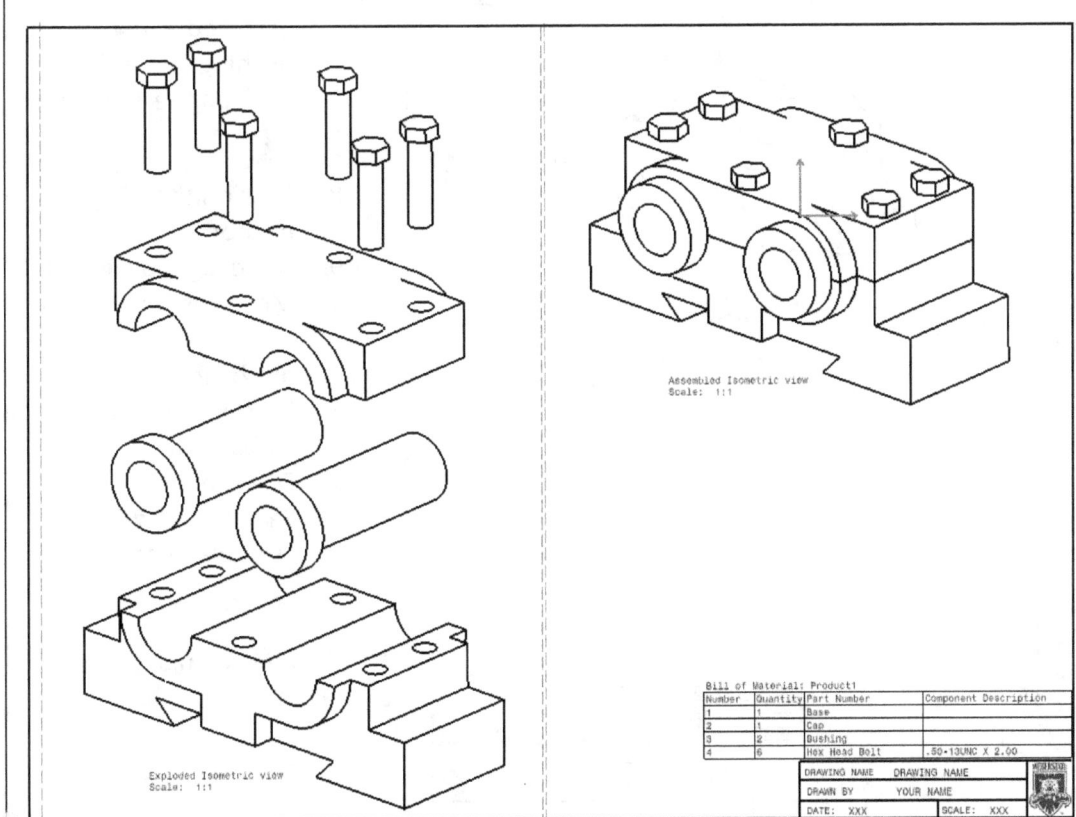

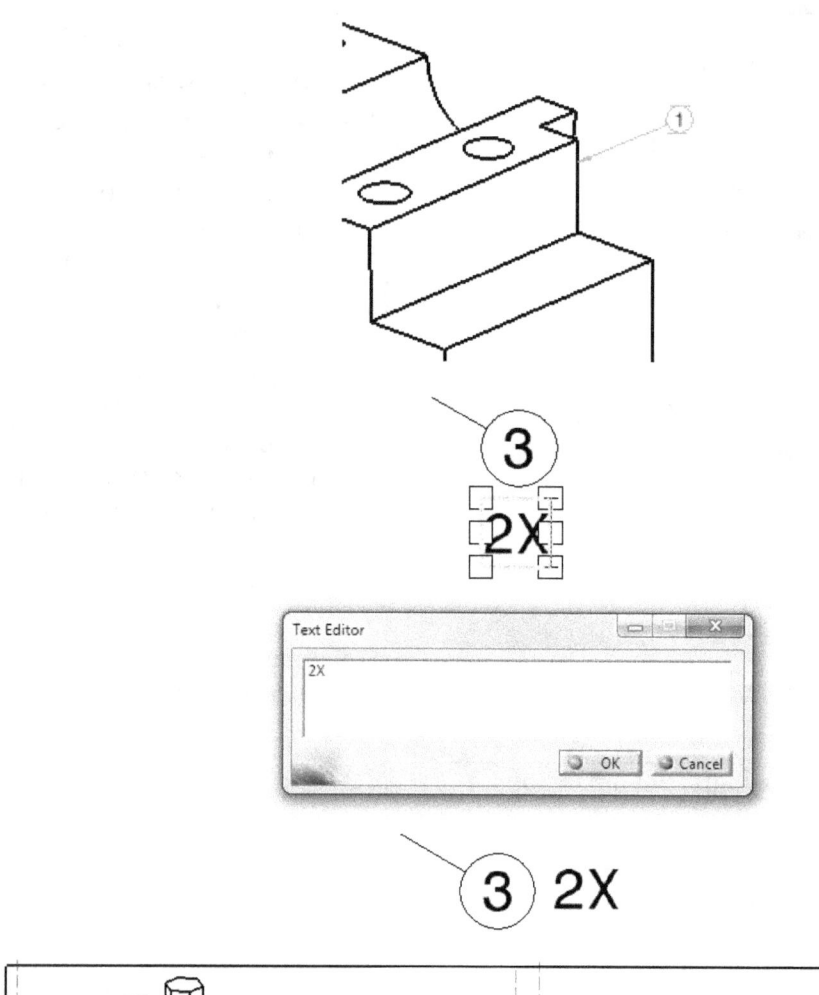

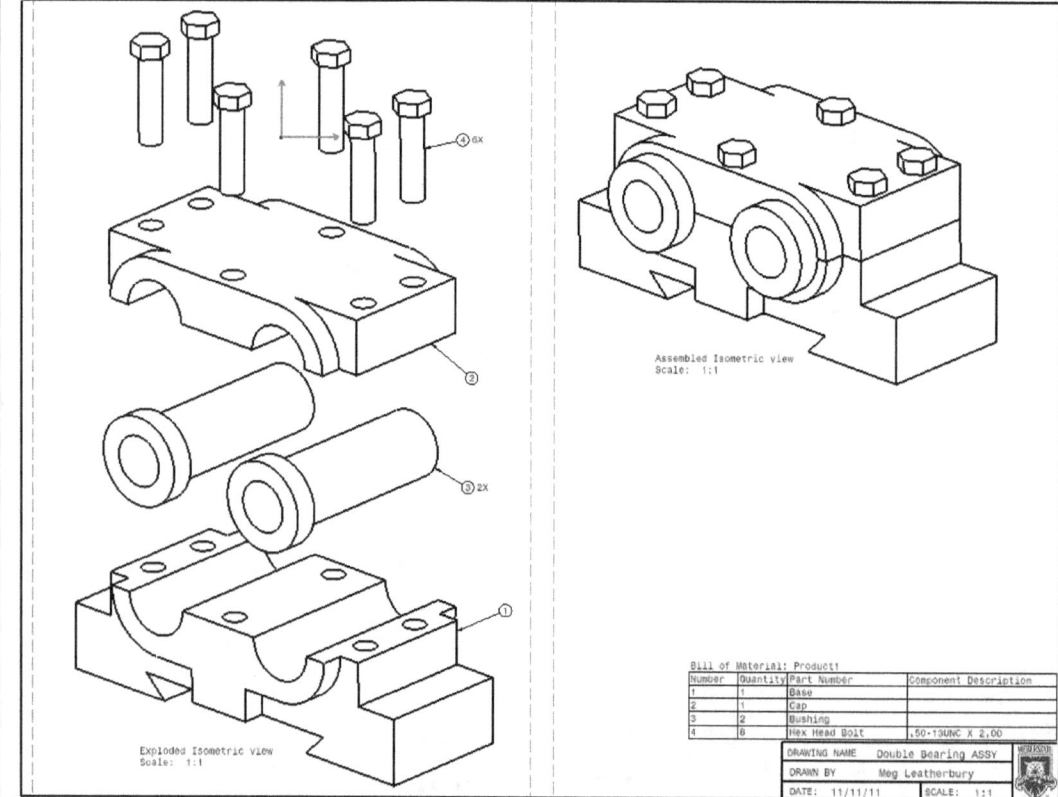

Now, let's add balloons on each part.

1. Activate the exploded Isometric View (double click the frame).
2. Under the Text icon **T**, click on the Balloon icon ⑥. (you can double-click so that it is activated until you are finished with that tool).
3. Click on an edge of the Base.
4. Click OK on the Balloon Creation box.
5. Repeat the same step on every part.

Cap: 2
Bushing :3
Bolt: 4

6. Let's add a text to the Bushing and Bolt.
7. Click on the Text icon **T**.
8. Click on the #3 balloon.
9. Type 2X and click OK.
10. Move the text next to the balloon as shown on the left.
11. Do the same thing to #4 balloon.
12. Edit the title block.
13. Save the drawing as "Double Bearing ASSY."

The drawing should be similar to the drawing on the left.

Assignment:

Create an assembly and drawing of a toy truck similar to below.

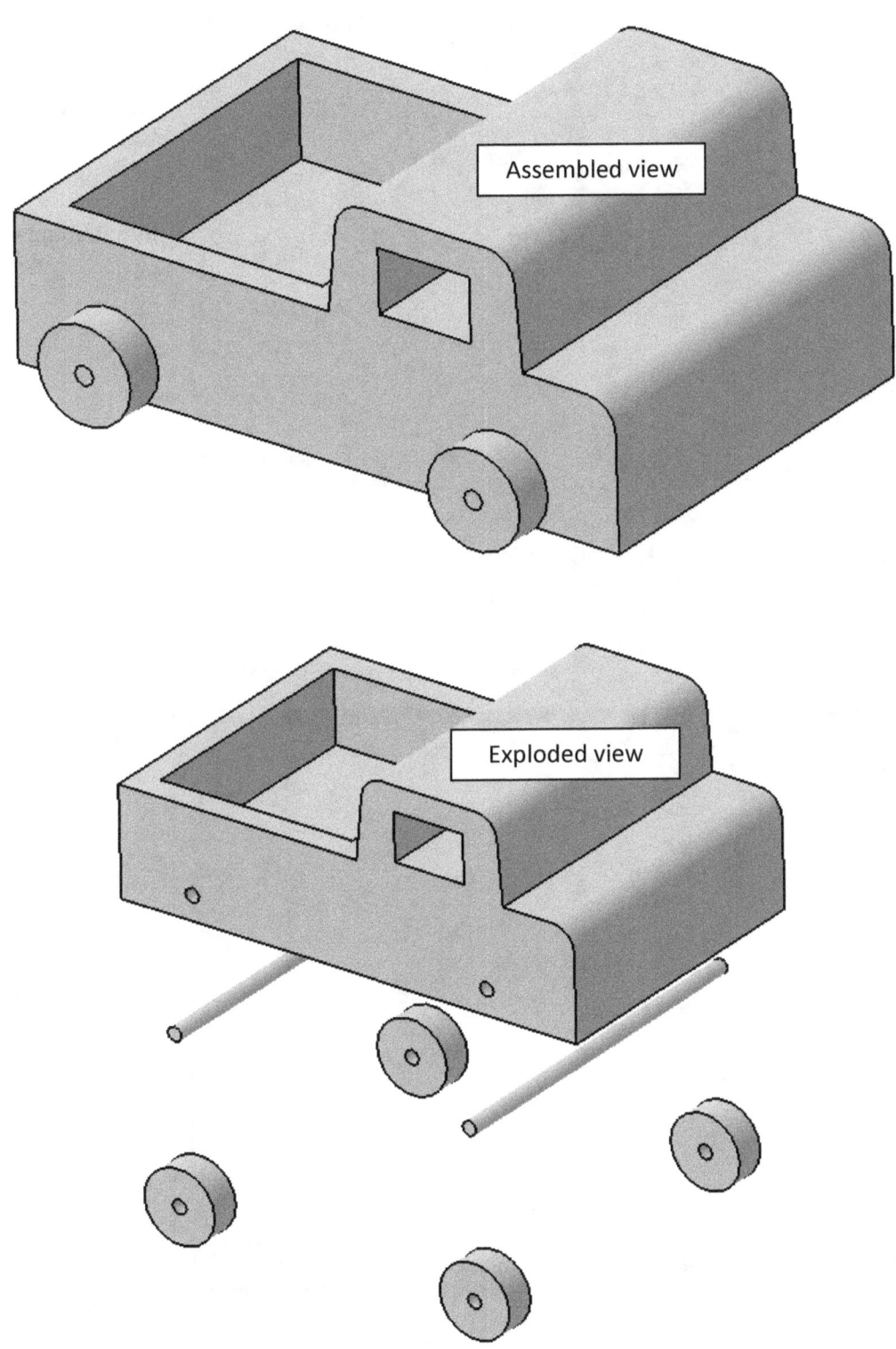

Chapter 6 – Generative Shape Design Workbench

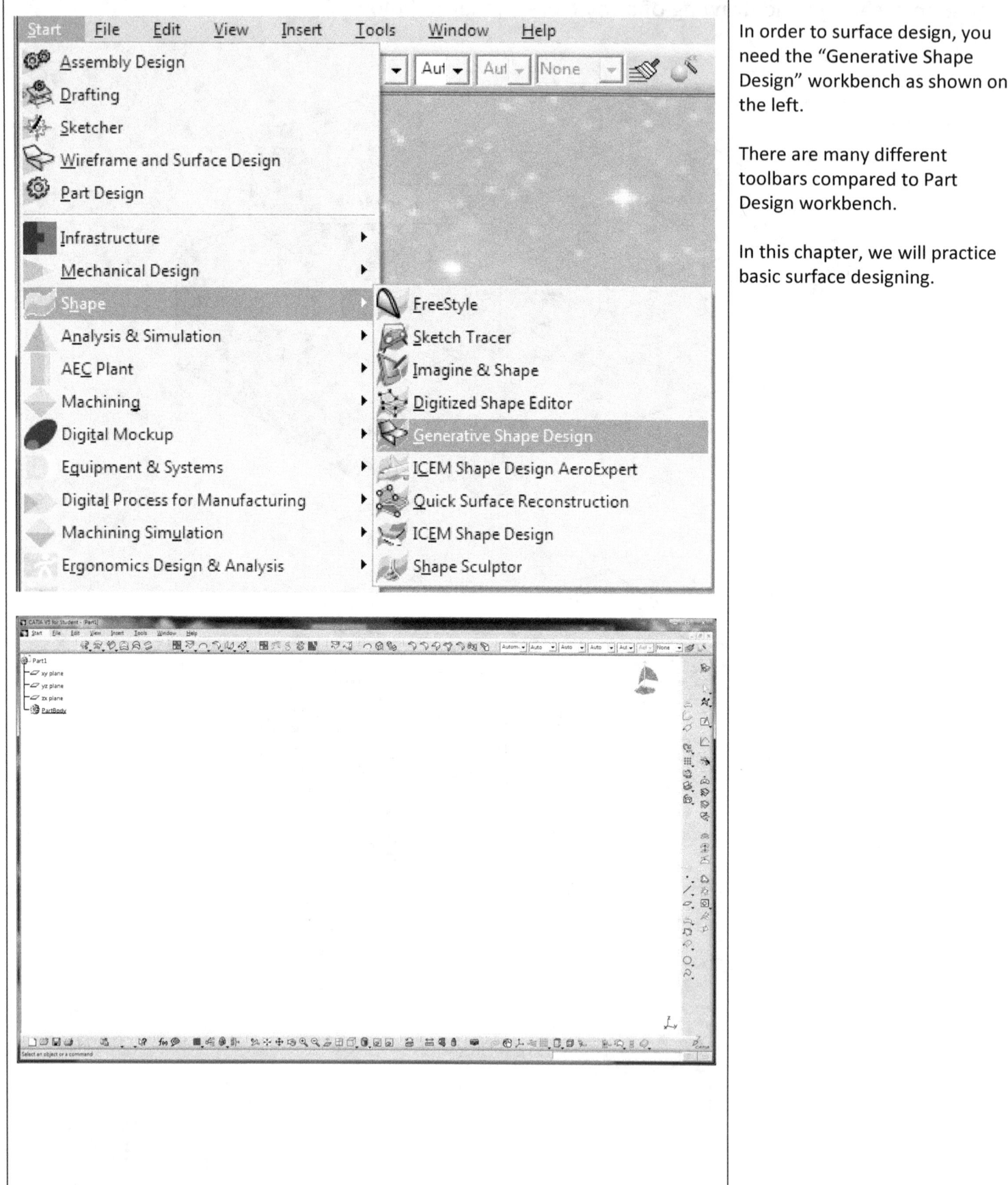

In order to surface design, you need the "Generative Shape Design" workbench as shown on the left.

There are many different toolbars compared to Part Design workbench.

In this chapter, we will practice basic surface designing.

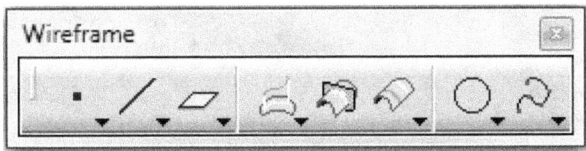

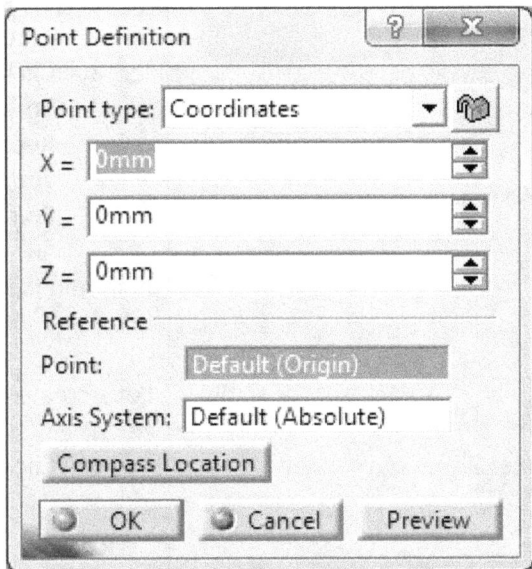

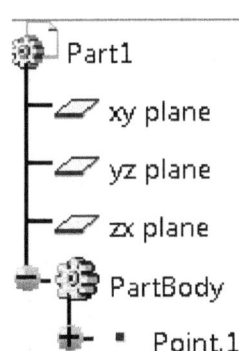

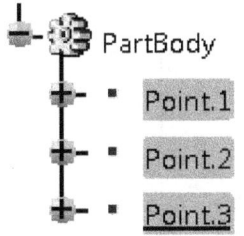

Let's create sheet metal from surface designing.

1. Open Generative Shape Design and set it up to inches.
2. Make sure you have the "Wireframe" toolbar – this is important in surface designing.
3. Click on the Point icon .
4. In the Point Definition box, enter 0,0,0 and click OK – this makes a point on the Origin as shown on the left.
5. Repeat the same step, create the following points.

   1,0,0
   0,0,2

   You should have three points as shown on the left.

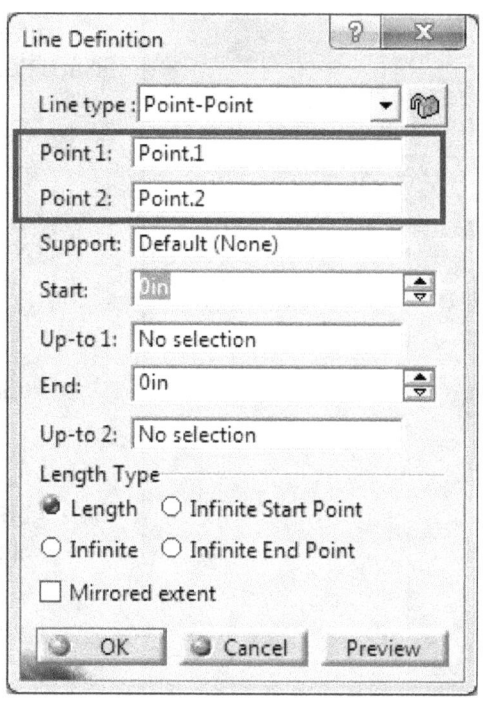

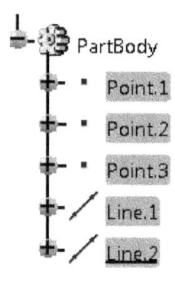

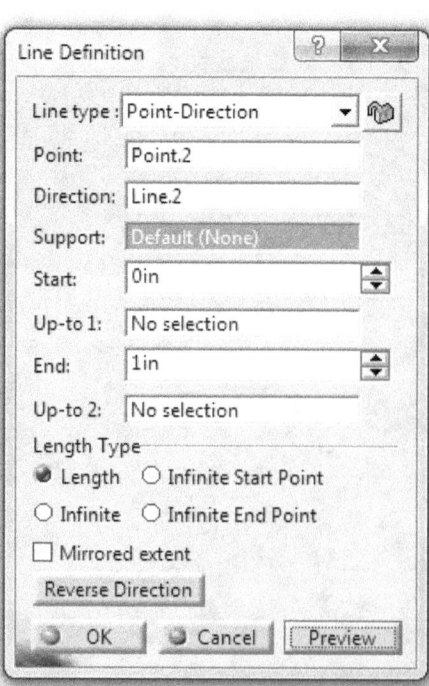

Let's connect those lines.

6. Click on the Line icon
7. Click Point.1 from the Tree for Point 1
8. Click Point.2 from the Tree for Point 2.
9. Repeat the same step and connect Point 1 and Point 3.

It should look as shown on the left.

Let's create more lines.

1. Click on the Line icon
2. Set up the Line Definition box as shown on the left.
3. Click OK.

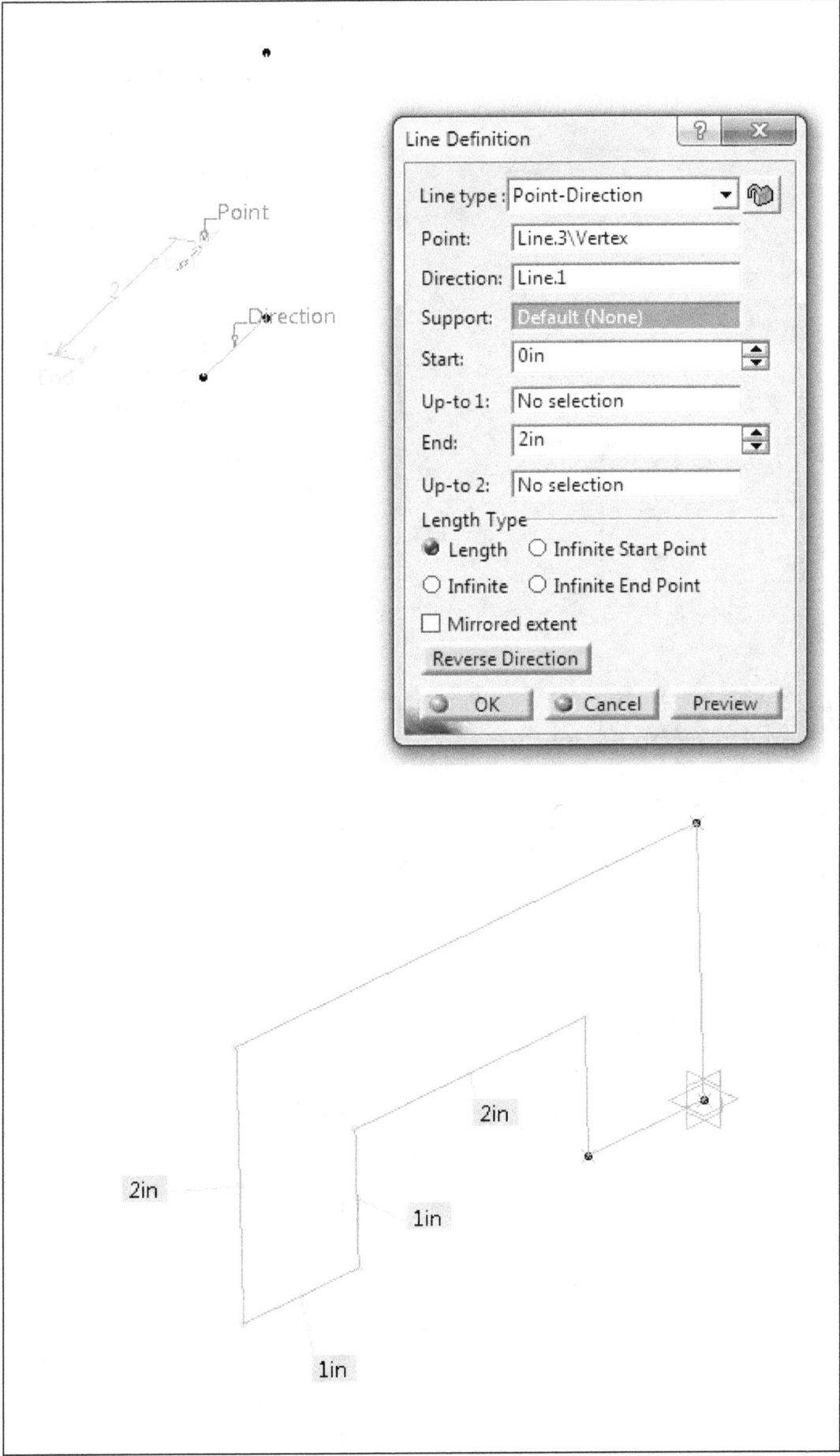

4. Repeat the same step and create a line as shown on the left.
5. Repeat the same step and create lines as shown on the left.

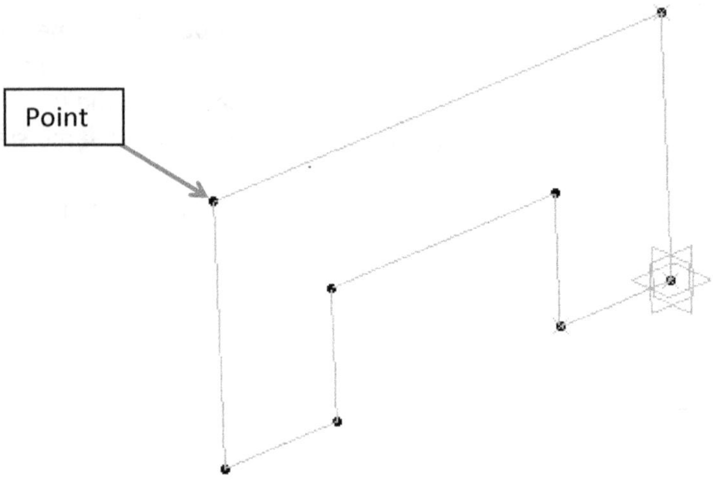

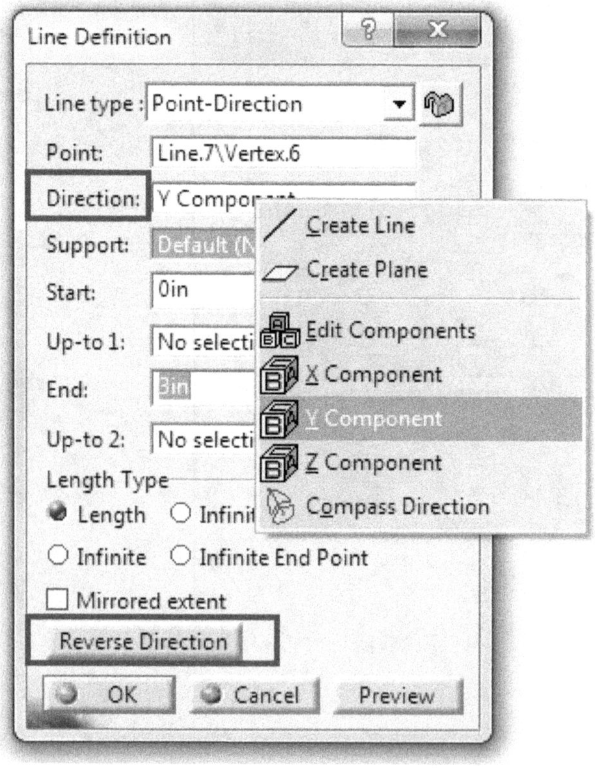

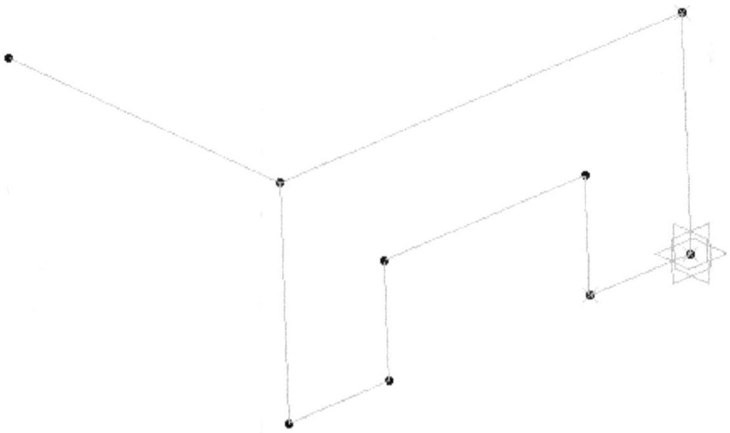

Lets' create more lines.

1. Click on the Line icon  .
2. Click the corner as shown on the left
3. In the Direction box, right click and select "Y Component."
4. The end value is 3 inches.
5. If the direction is facing the wrong way, click "Reverse Direction."

It should be similar to the sketch on the left.

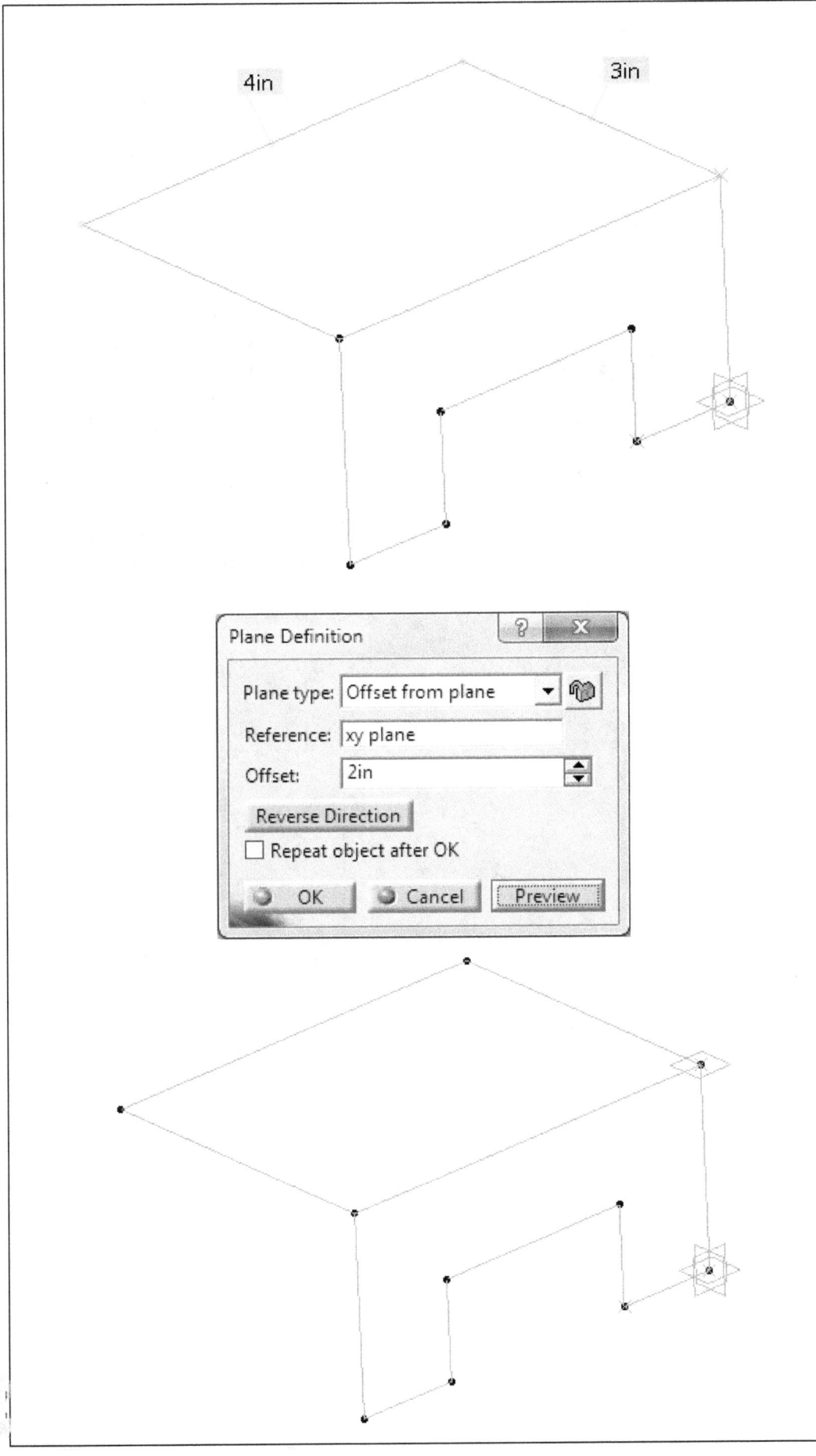

6. Repeat those steps, and create the lines as shown on the left.

Add corners.

1. First, offset a plane by clicking on the Plane icon.
2. Plane type is "Offset from plane."
3. Select XY plane from the Tree for Reference.
4. Enter 2 inch for Offset value.

The offset plane should be similar to the sketch on the left.

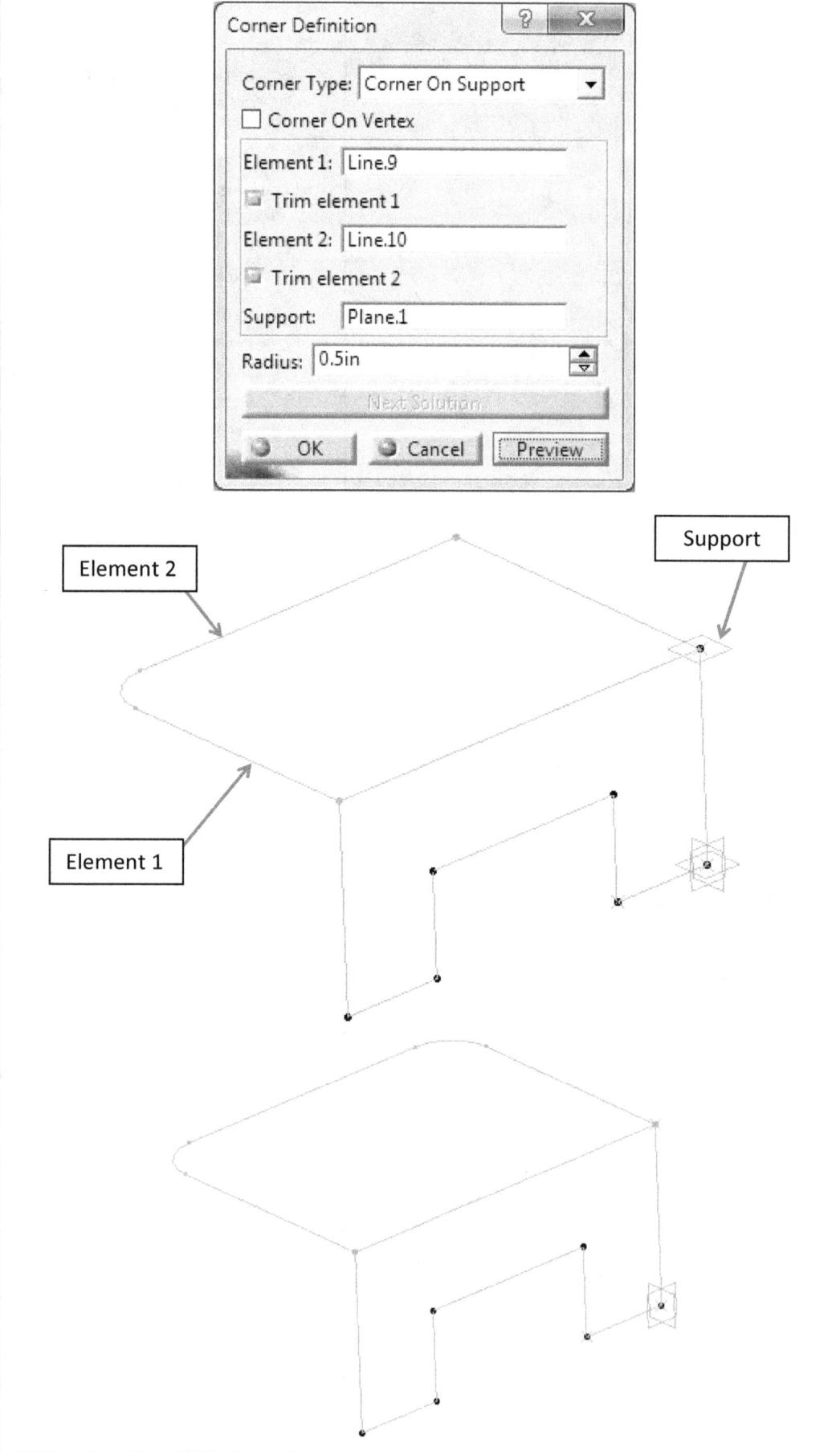

5. Click on the Corner icon under the Circle icon .
6. Set the Corner Definition box as shown on the left.
7. Make sure "Trim element 1" and "Trim element 2" are activated.
8. Repeat the step 7 and do the same thing on the opposite side.

The result be similar to the sketch on the left.

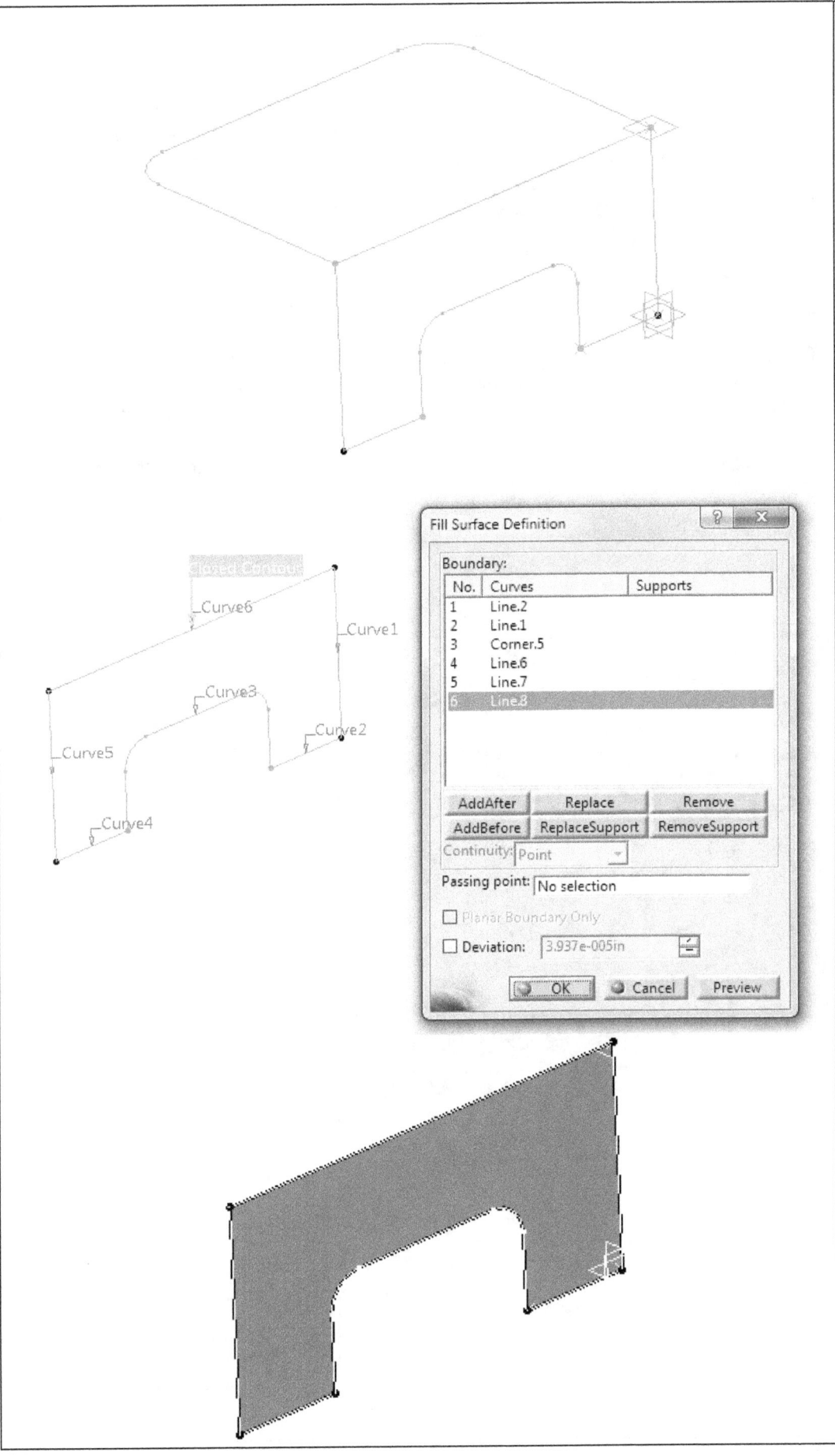

Using the same method, create corners on the small opening.

Use XZ plane for Support
Radius: 0.3

The result be similar to the sketch on the left.

Let's add surfaces to these elements.

1. Click on the Fill icon.
2. Click each line as shown on the left – select lines right next to each other. Do not skip any.
3. Click OK.

The result should look like on the left.

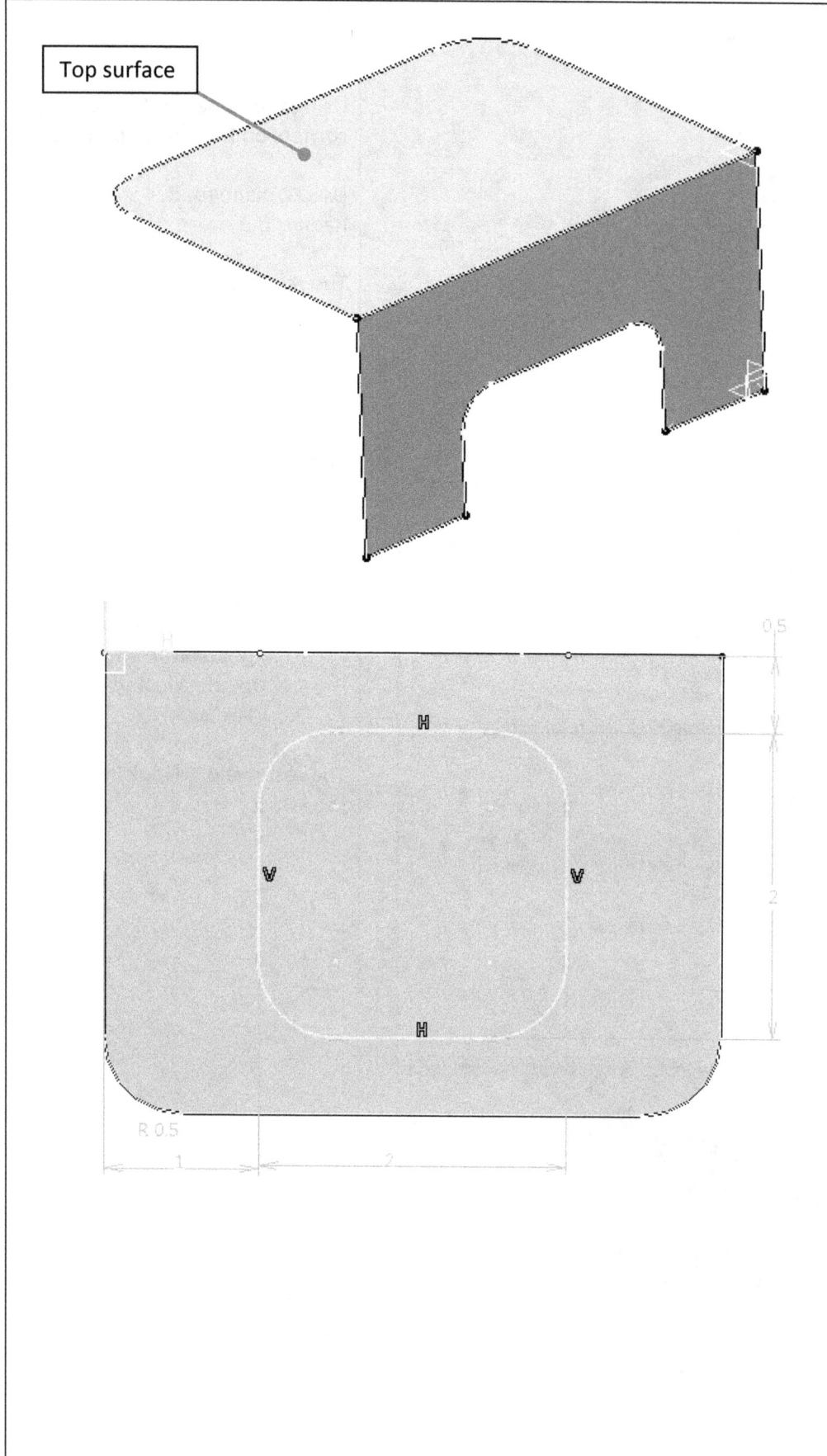

4. Repeat the same step and create a surface for the top. The result should look like on the left.

Let's add an opening on the top surface.

1. Click on the sketch icon and select the top surface.
2. Sketch as shown on the left.
3. Exit the Sketcher.

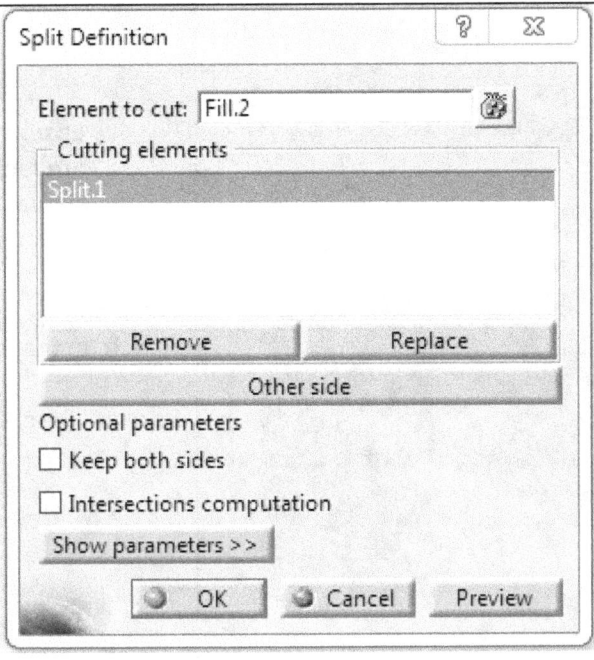

4. Click on the Split icon .
5. Select the top surface for "Element to cut."
6. Select the sketch you just created for the "Cutting elements."

The result should look like the left.

Add fillet on the edge.

1. Click on the Shape Fillet icon .
2. Click the top surface and the front surface.
3. Enter 0.5 for the Radius.

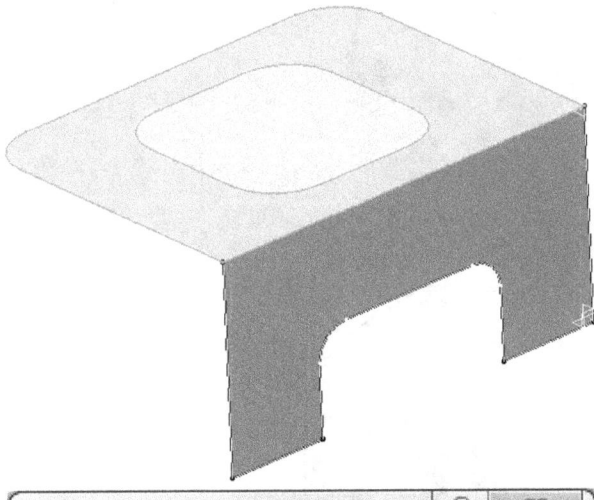

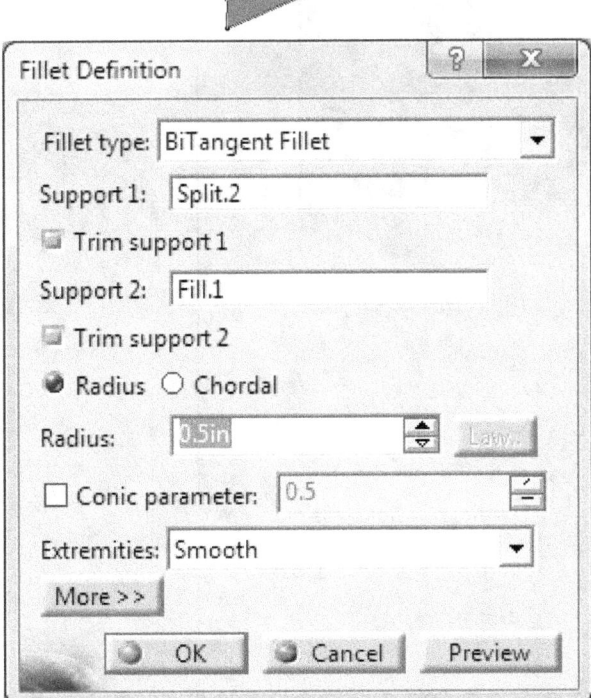

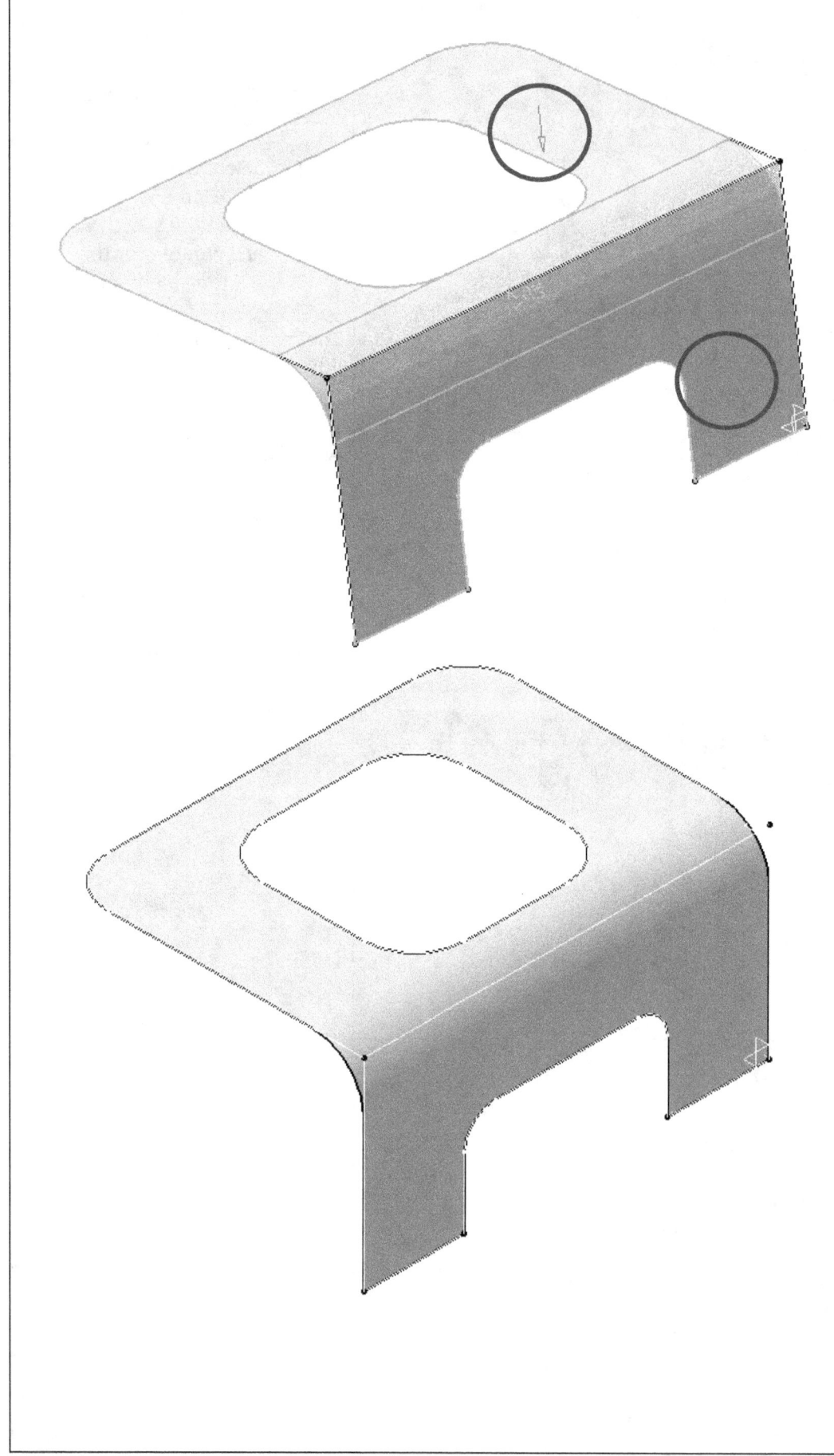

4. Ensure the arrowheads are facing inboard as shown on the left.
5. Click OK.

The result be similar to the sketch on the left.

Save as "Bracket."

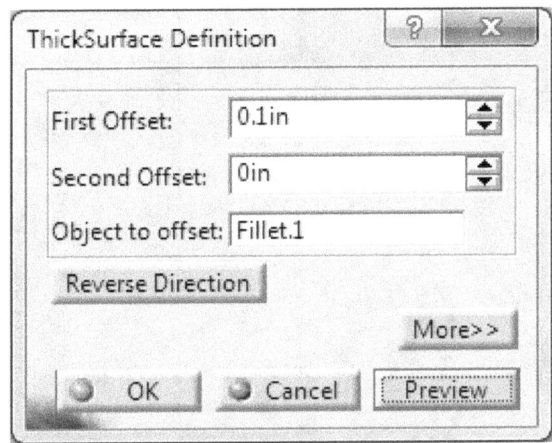

Let's add solid thickness and material.

1. With the Bracket open, go to Start, Mechanical Design, and Part Design.
2. Click on the Thick Surface icon  under the Close Surface icon.
3. Enter 0.1 inch in the First Offset.
4. Click the Fillet.1 on the Tree for the Object to offset.
5. Ensure the arrowheads are facing inboard as shown on the left.
6. Click OK.
7. Hide all the elements except the solid on the Tree.
8. Apply the material of your choice to the solid.

It should look similar to the sketch on the left.

Save early and often.

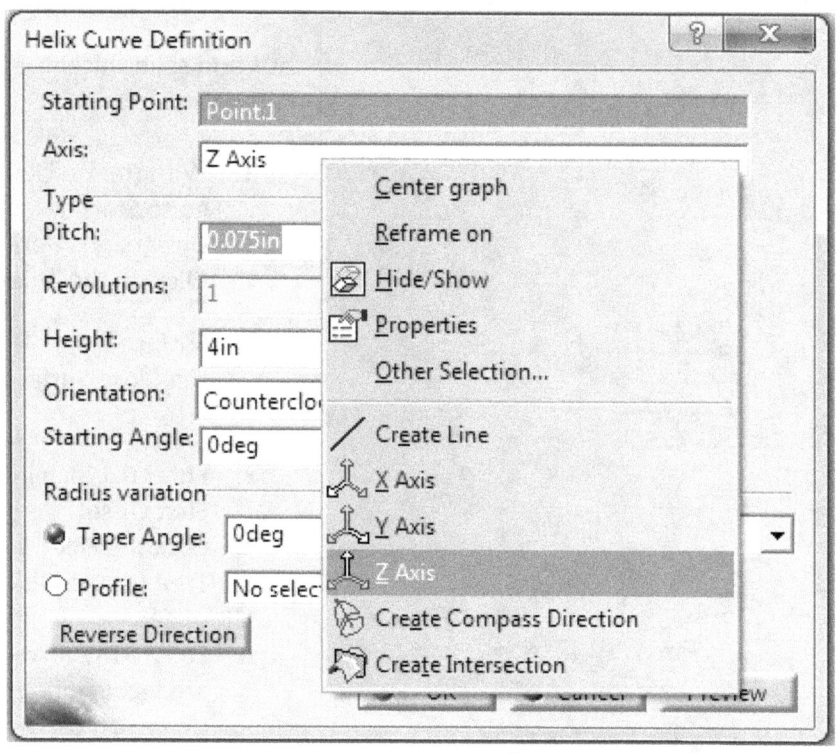

Let's practice Helix .

1. Open a new Generative Shape Design workbench.
2. Set it to inches.
3. Create a starting point at 0, 1.5, 0.
4. Select the Helix icon under the Spline icon.
5. Set up the Helix Curve Definition box as shown on the left. Right click on the Axis and select Z axis.
6. Enter 0.075 in the Pitch.
7. Enter 4 in the Height.
8. Click OK.
9. Create these points.
   0, 1.75, 0
   0, 1.5, .025
   0, 1.75, .025
10. Create lines connecting the four points as shown on the left.
11. Click on the Join icon and connect those four lines.

Notice Join.1 was created on the Tree.

Go to the Part Design workbench.

76

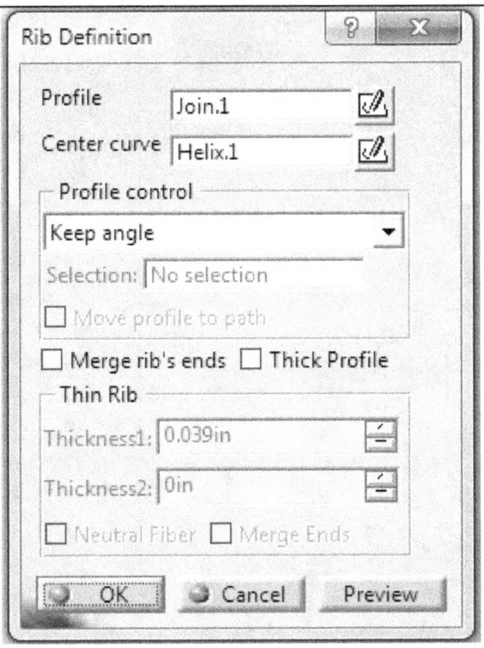

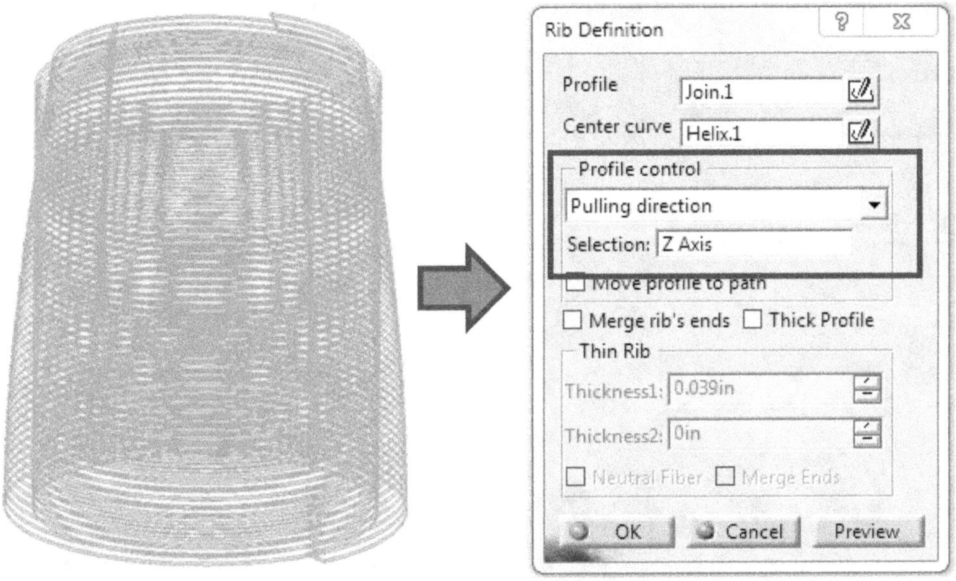

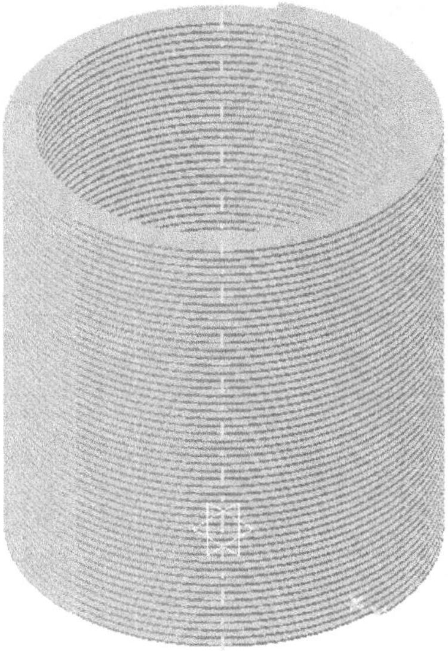

12. Click on the Rib icon  .
13. Select Join.1 for the Profile.
14. Select Helix.1 for the Center curve.

If the review looks similar to the sketch on the left, edit the Profile control as shown on the left.

The result should look similar to the sketch on the left.

Save as "Helix."

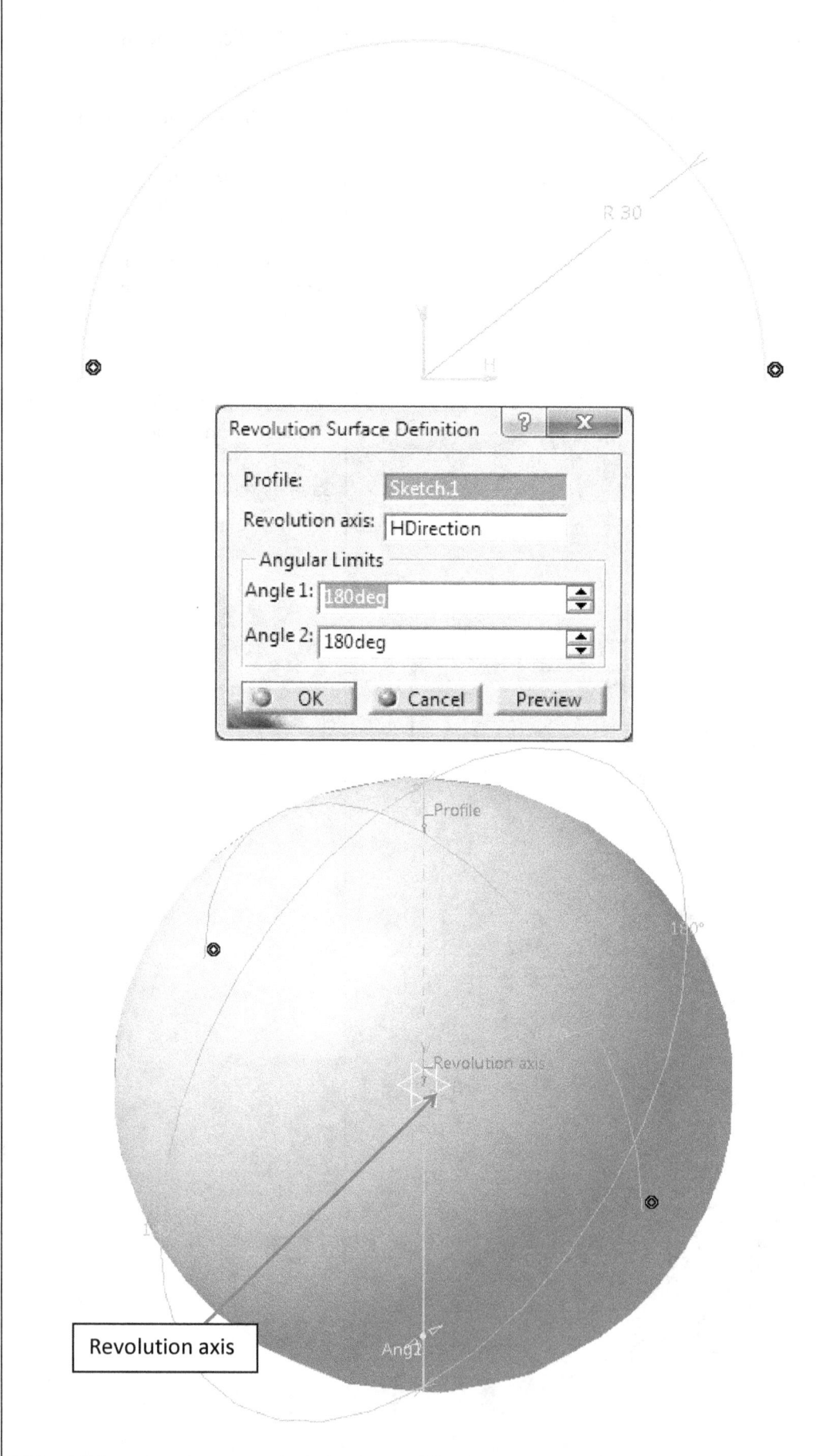

1. Select the YZ plane and create a sketch of an arc with R30 mm.
2. Exit the sketcher.
3. Click on the Revolve icon under the Extrude icon.
4. Set up the definition box as shown on the left.
5. Click OK.

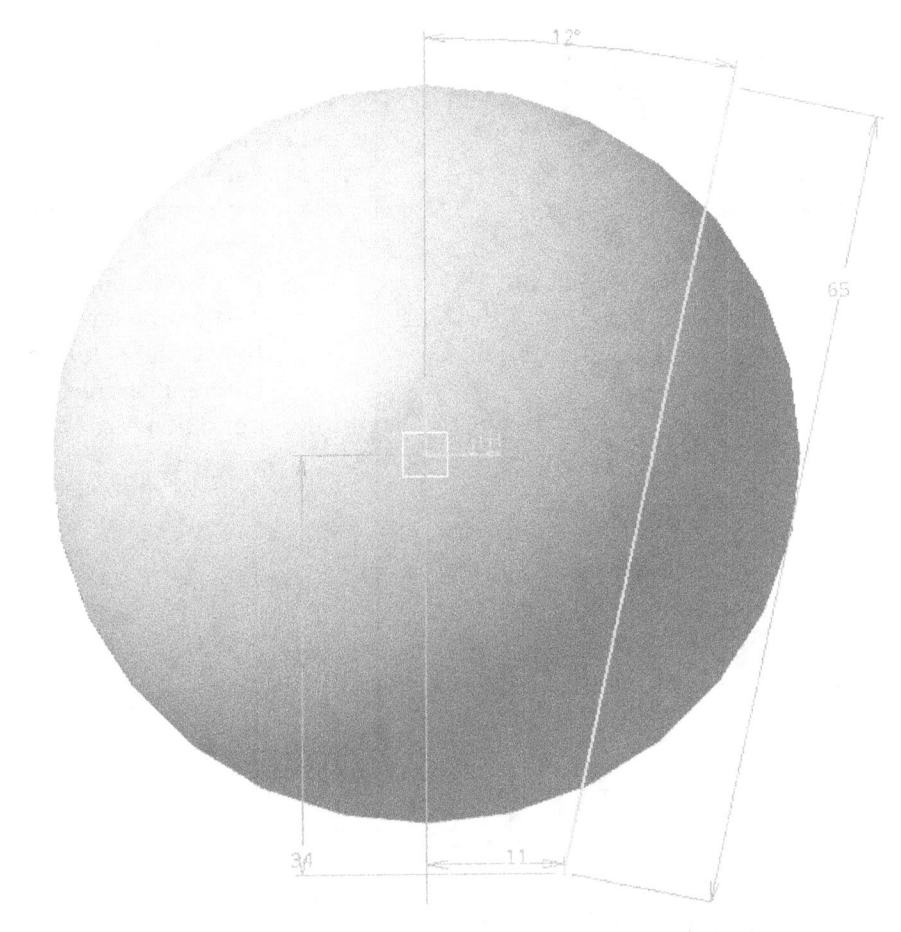

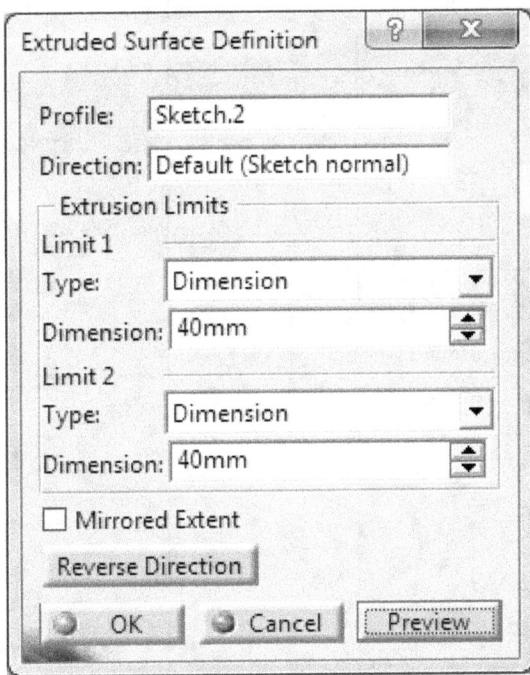

6. Sketch as shown on the left on the YZ plane.
7. Exit the Sketcher.
8. Click on the Extrude icon.
9. Set up the definition box as shown on the left.
10. Click OK.

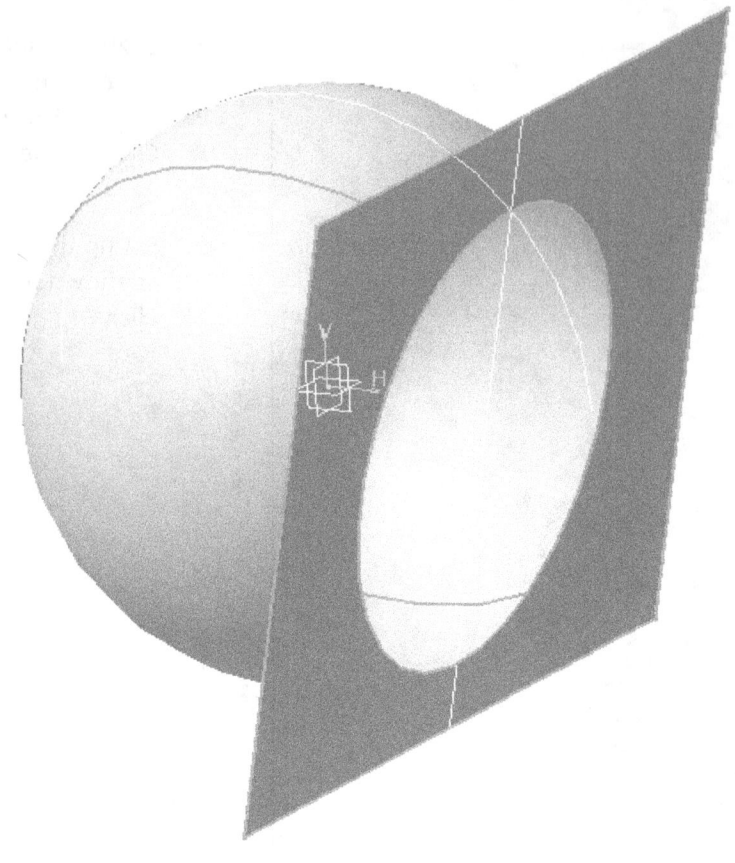

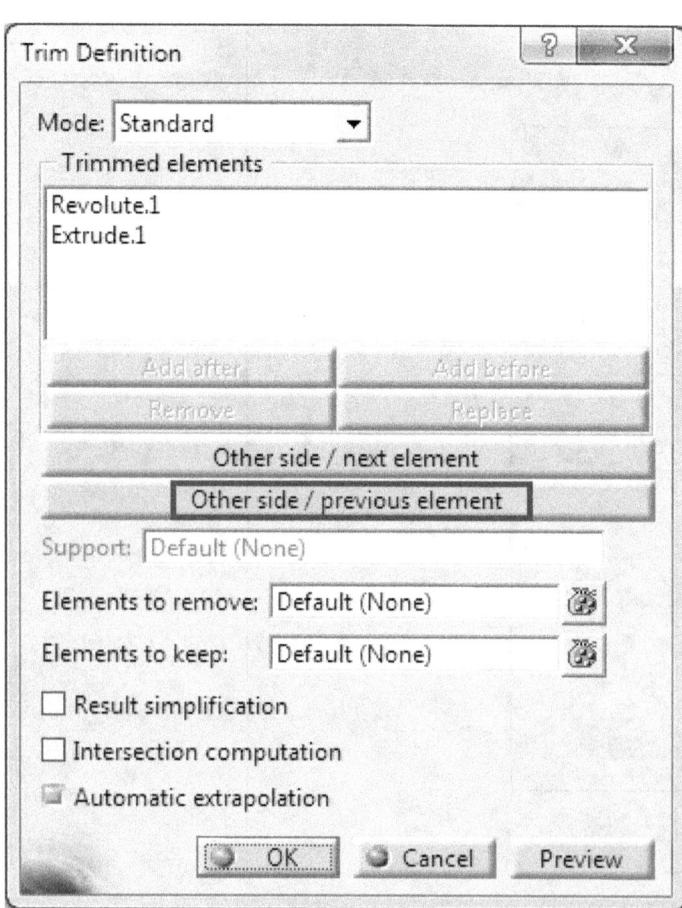

11. Click on the Trim icon.
12. First click the sphere.
13. Second, click the extrude that you just create.
14. If it looks similar to the part on the left, click on the "Other side/previous element" in the definition box.

80

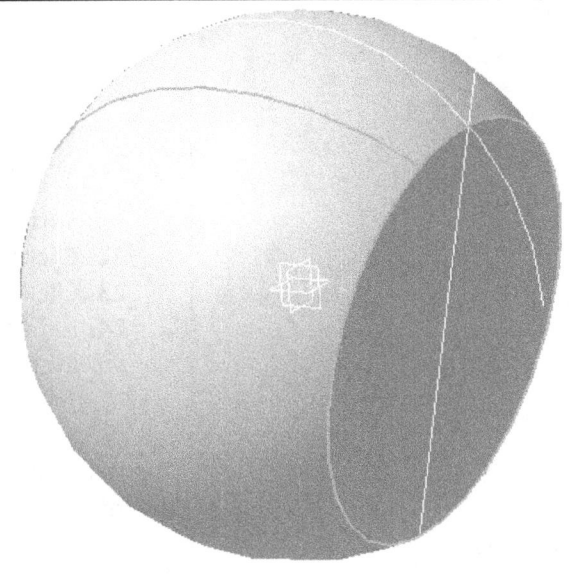

15. If it looks similar to the part on the left, click OK.
16. Click on the Edge Fillet icon.
17. Set up the definition box as shown on the left.
18. Click on the edge.
19. Click OK.

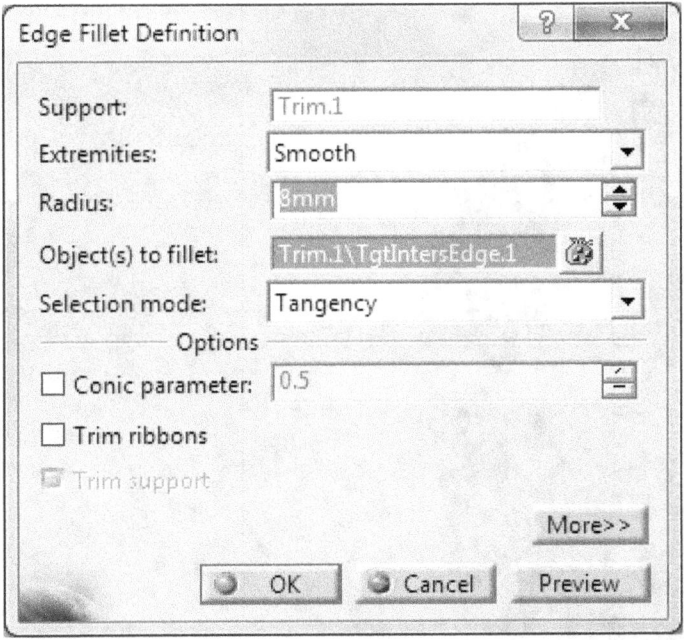

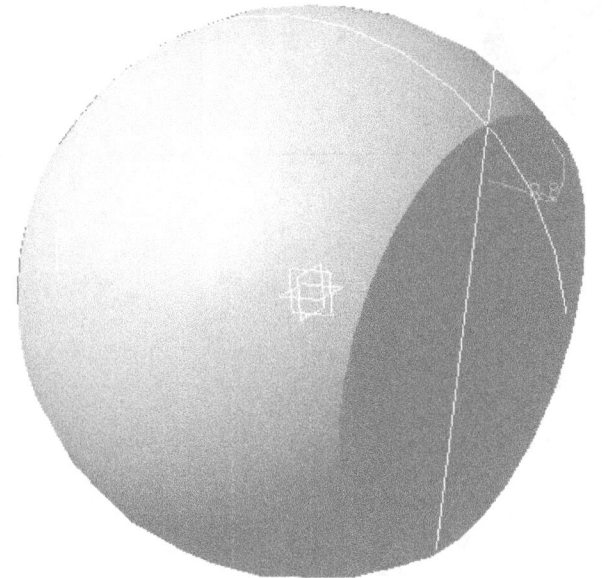

81

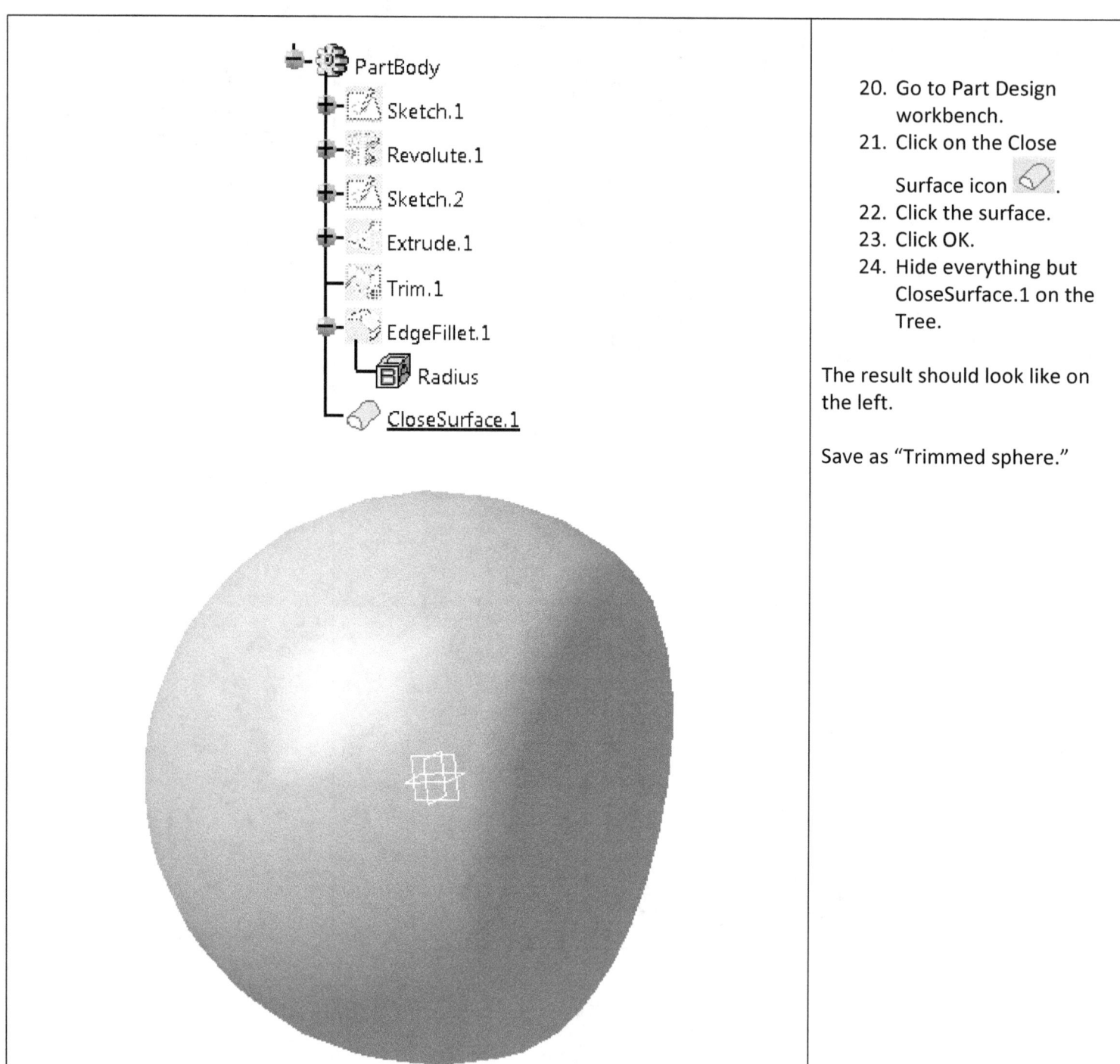

20. Go to Part Design workbench.
21. Click on the Close Surface icon .
22. Click the surface.
23. Click OK.
24. Hide everything but CloseSurface.1 on the Tree.

The result should look like on the left.

Save as "Trimmed sphere."

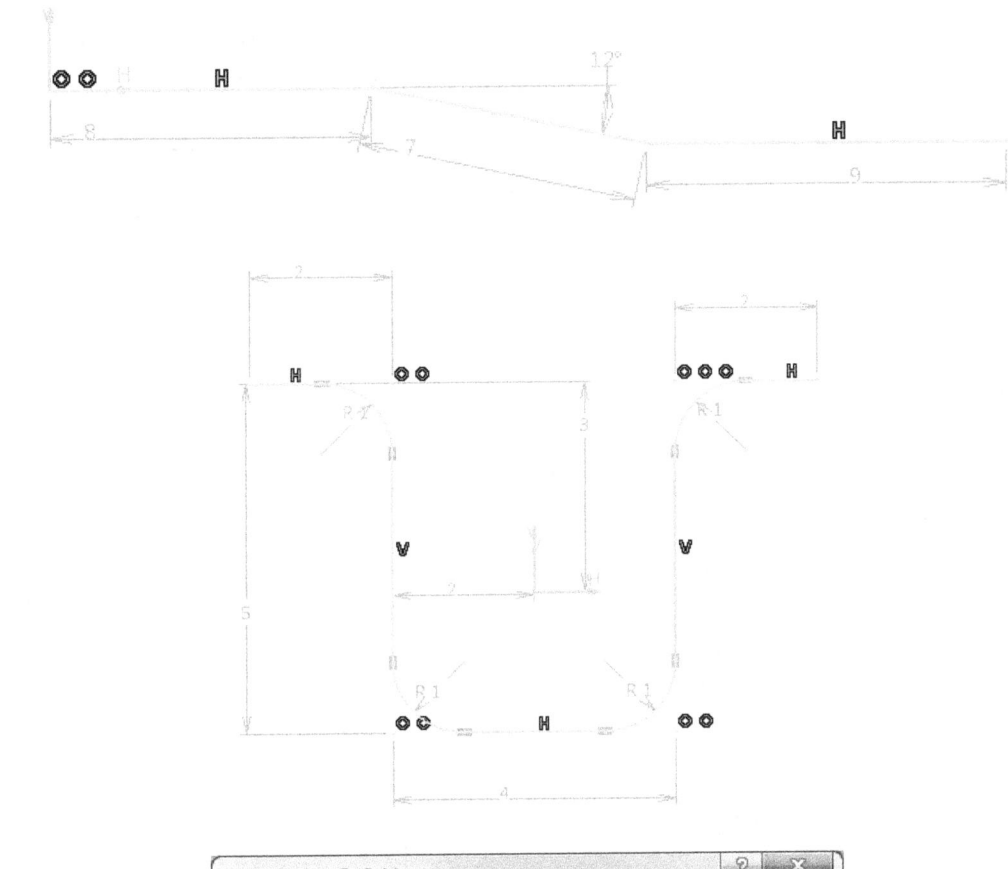

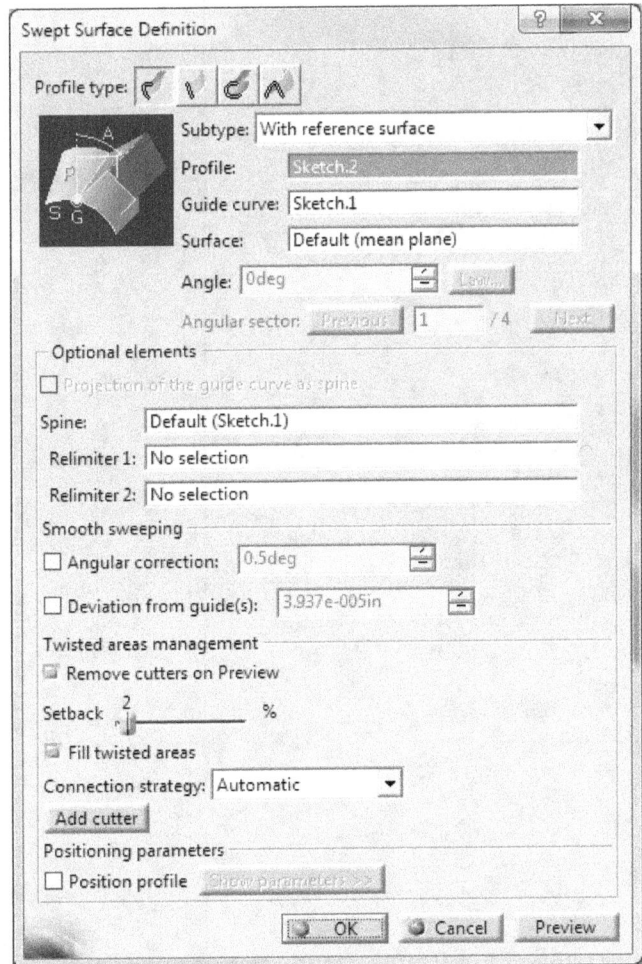

1. Open a new Generative Shape Design.
2. Set up in inches.
3. Sketch as shown on the left on the YZ plane.
4. Get out of the Sketcher.
5. Sketch as shown on the left on the ZX plane.
6. Get out of the Sketcher.
7. Click on the Sweep icon.
8. Set up the definition box as shown on the left.
9. Click OK.

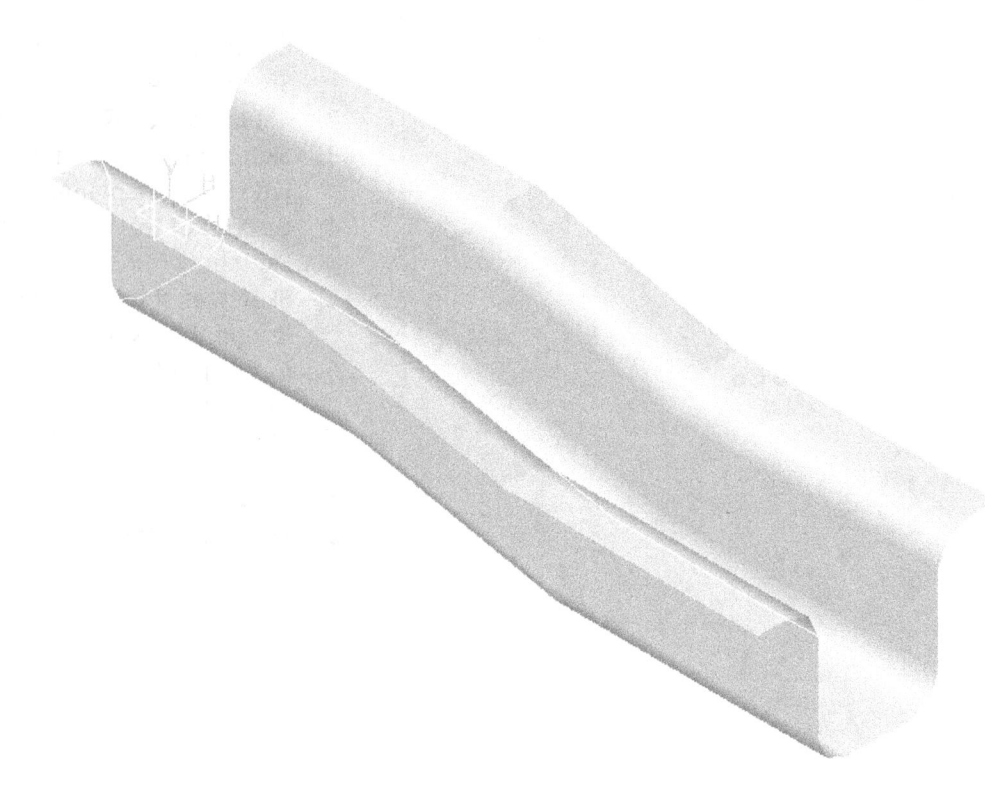

The result should be similar to the part on the left.

Let's extend on of the flange.

1. Click on the Extrapolate icon.
2. On the definition box, right click on the Boundary and select Create Join.

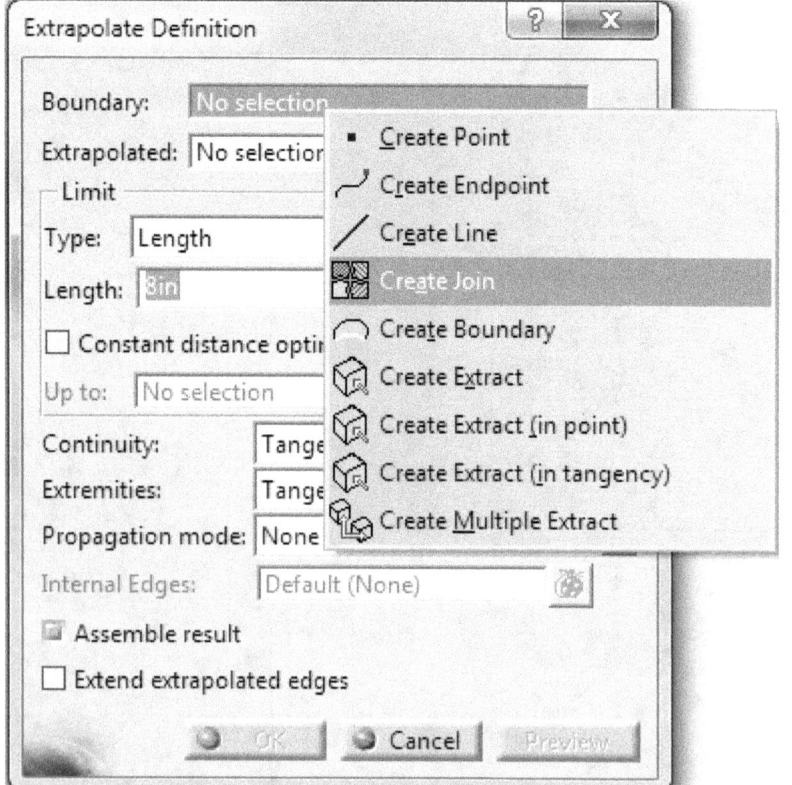

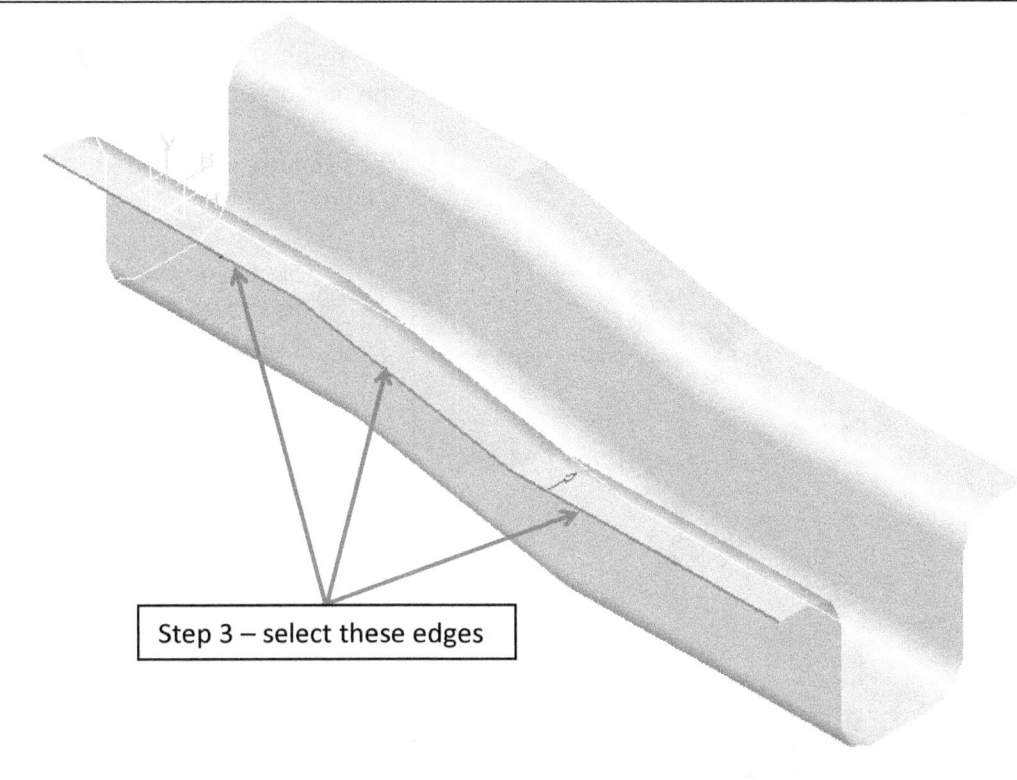

Step 3 – select these edges

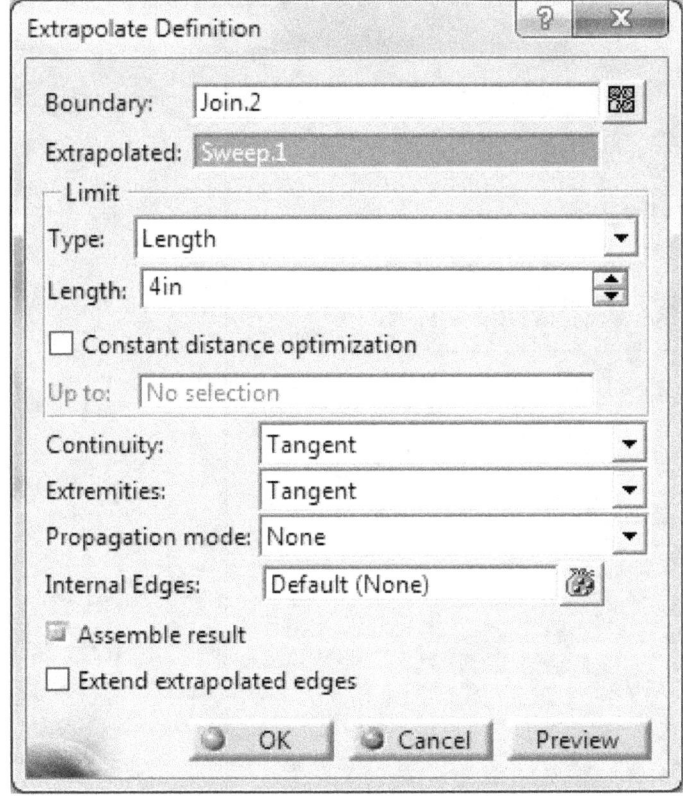

3. Click three edges of the flange as shown on the left.
4. Click OK.
5. Set up the definition box as shown on the left.
6. Click OK.

The result should be similar to the part on the left.

Let's shape the extended flange.

1. Sketch on the XY plane as shown on the left.
2. Exit the Sketcher.

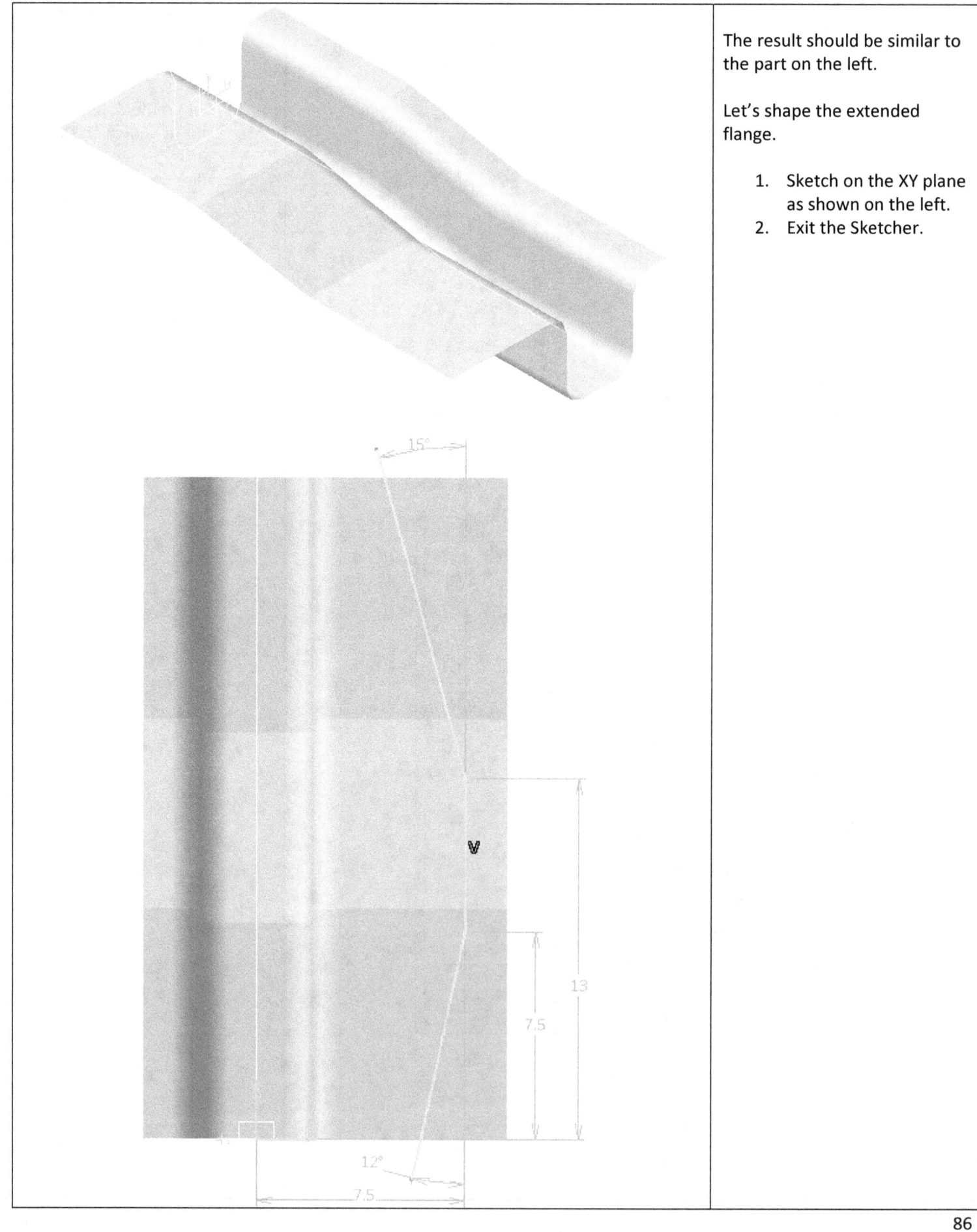

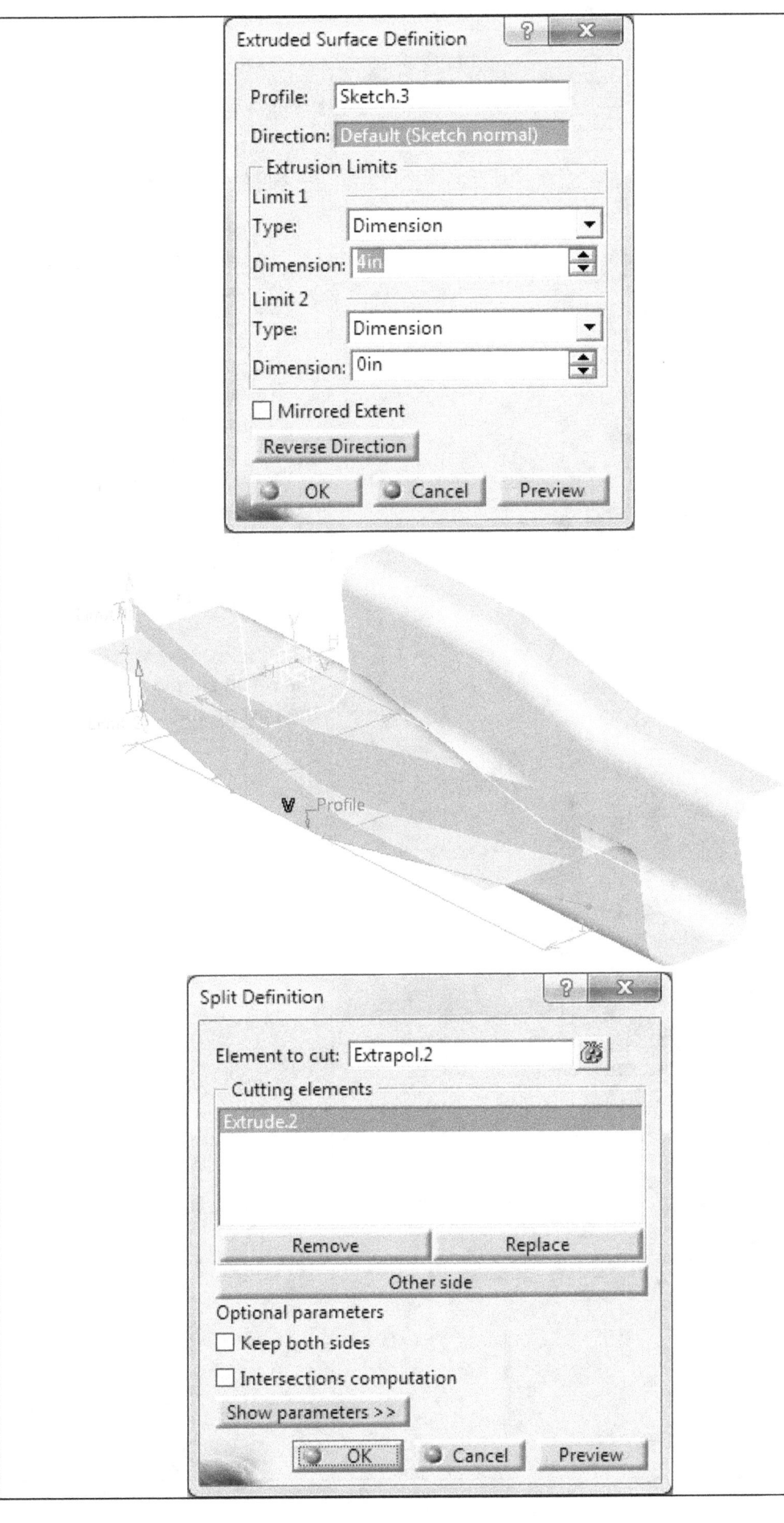

3. Click on the Extrude icon .
4. Set up the definition box as shown on the left.
5. Make sure the extrude goes through the flange.
6. Click OK.
7. Click on the Split icon .
8. Set up the Split Definition box as shown on the left.

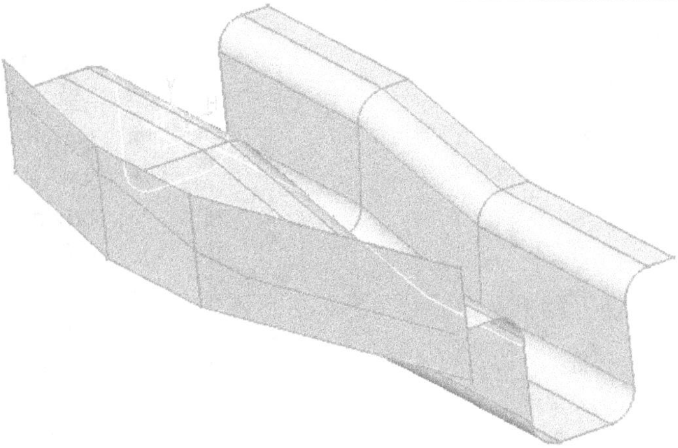

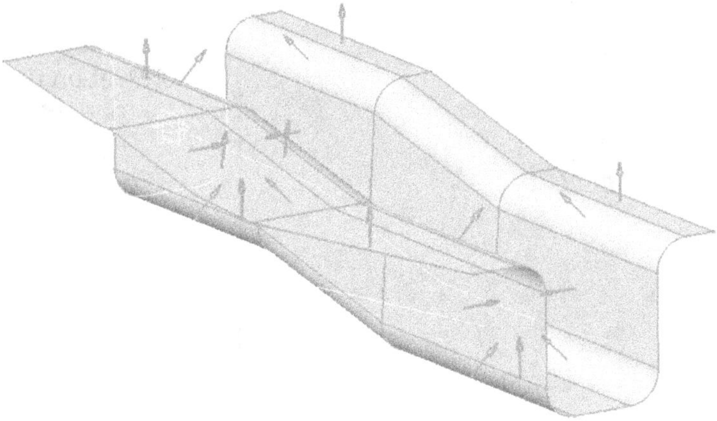

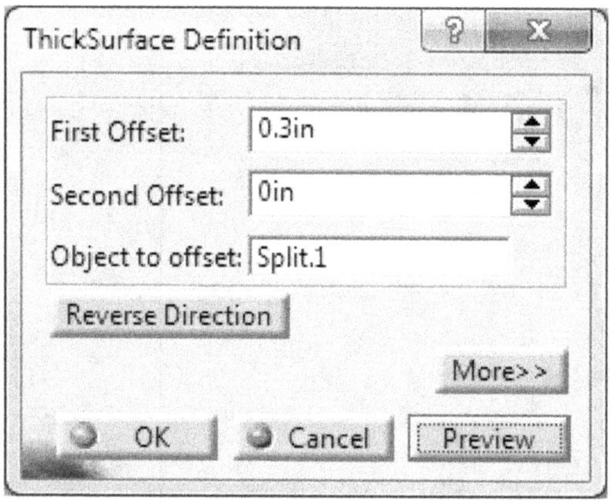

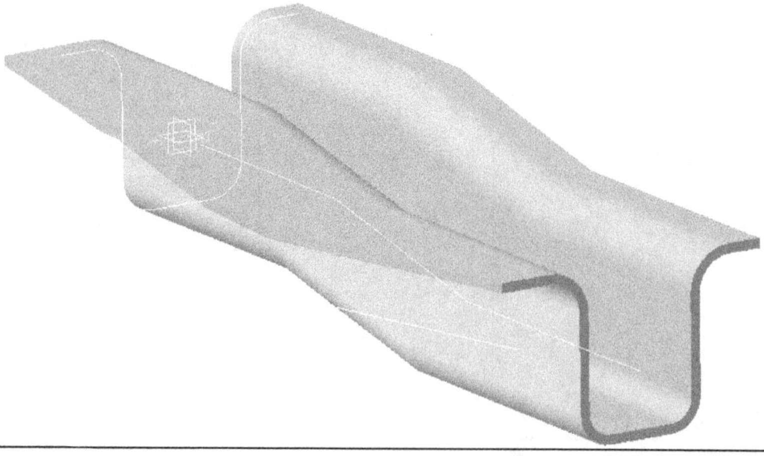

The flange was cut to the extrude as shown on the left.

9. Hide the Extrude.
10. Go to the Part Design workbench.
11. Click on the Thick Surface icon .
12. Make sure the direction of arrows are facing upward as shown on the left.
13. Set up the definition box as shown on the left.
14. Click OK.

The result should be similar to the part on the left.

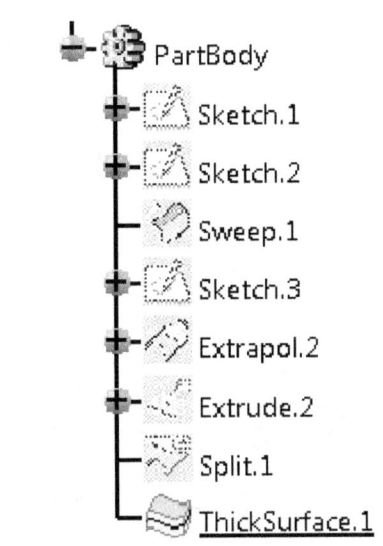

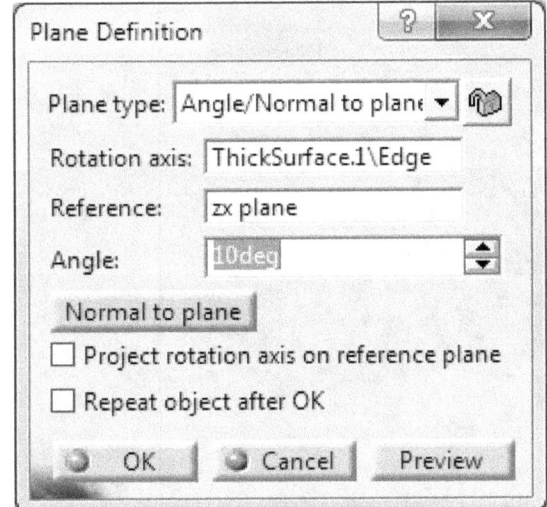

15. Hide all the elements except the thickness on the Tree.
16. Click on the Plane icon .
17. Set up the definition box as shown on the left.
18. Click OK.

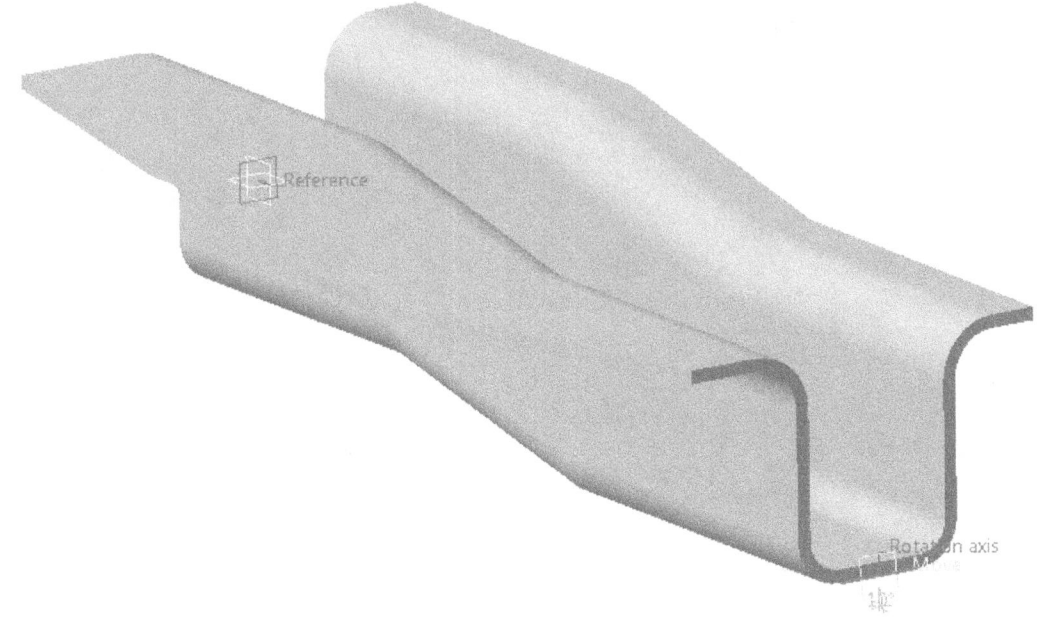

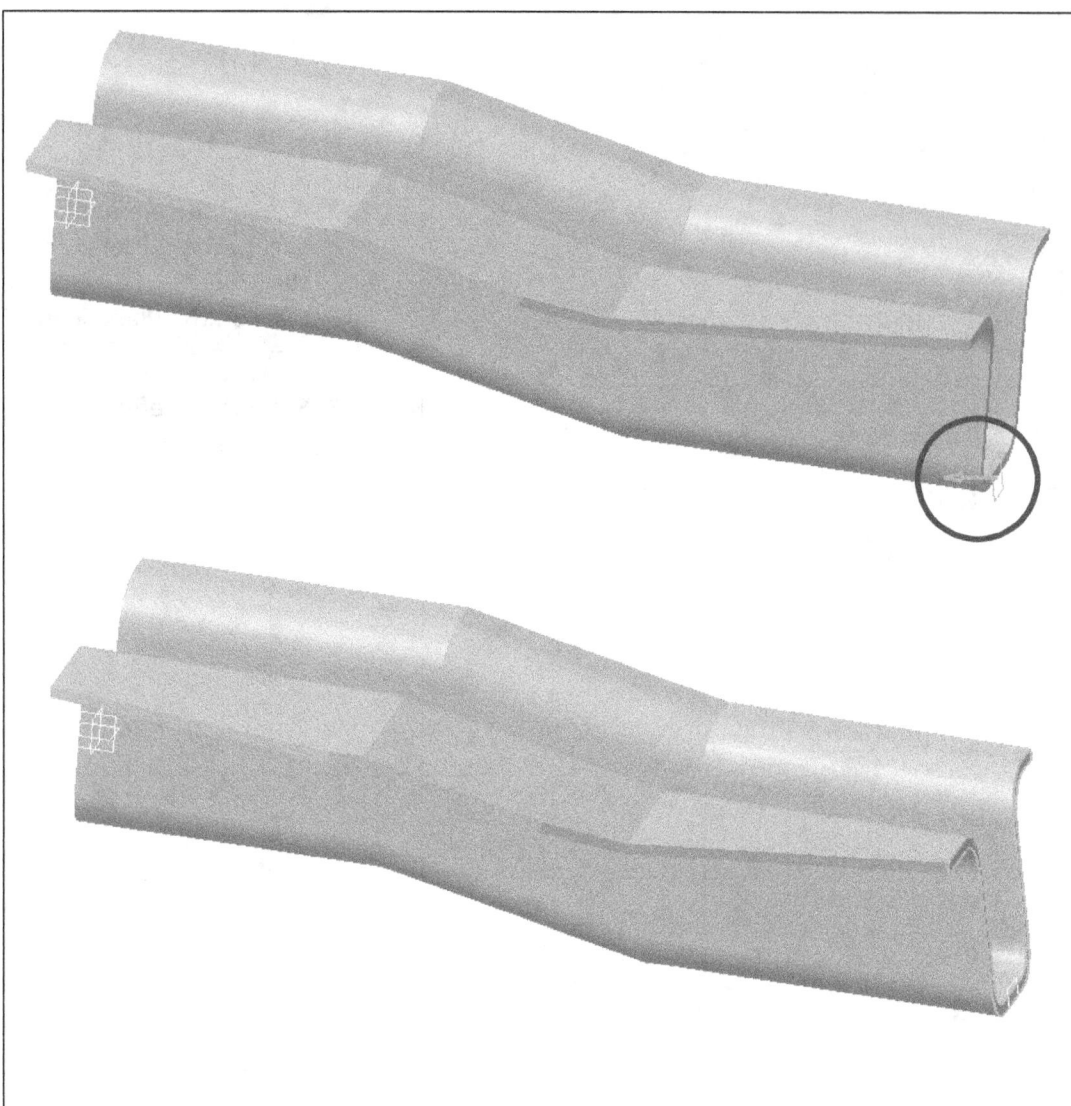

19. Click on the Split icon .
20. Select the plane you just created for the Splitting Element.
21. Ensure the direction of the arrow is correct.
22. Click OK.

The result should be similar to the part on the left.

Save as "Extruded bracket."

Chapter 7 – Rendering Workbench

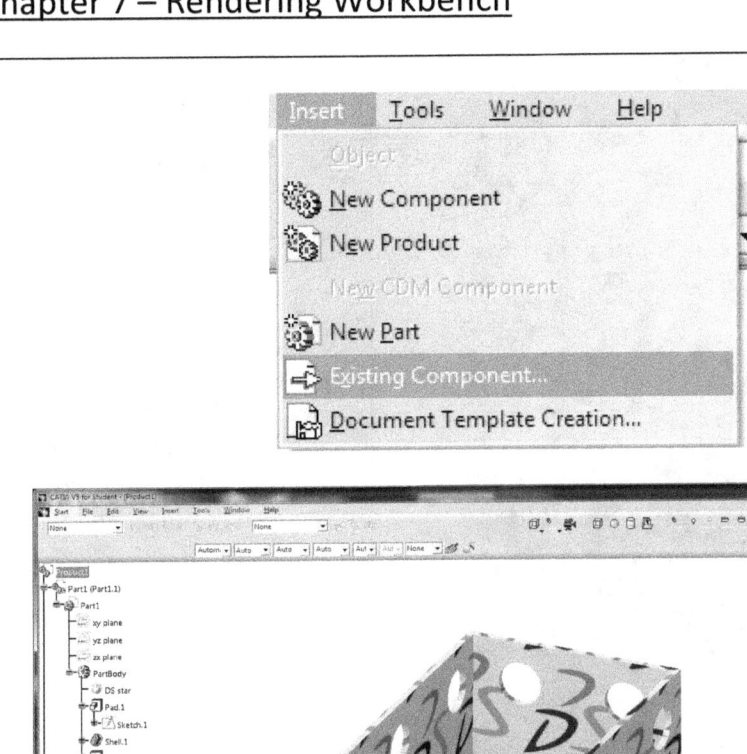

1. Go to Start, Infrastructure, and Photo Studio.
2. Go to Insert, and click "Existing Component."
3. Insert Chapter 3 "Shaded."

The "Create Environment" toolbar will be needed for this chapter.

4. Click the Create Box Environment icon ⬛. This creates a box around the part.

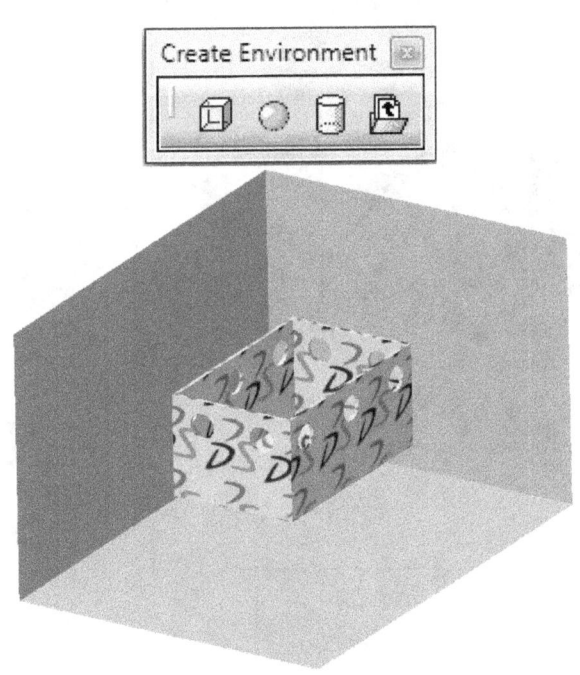

91

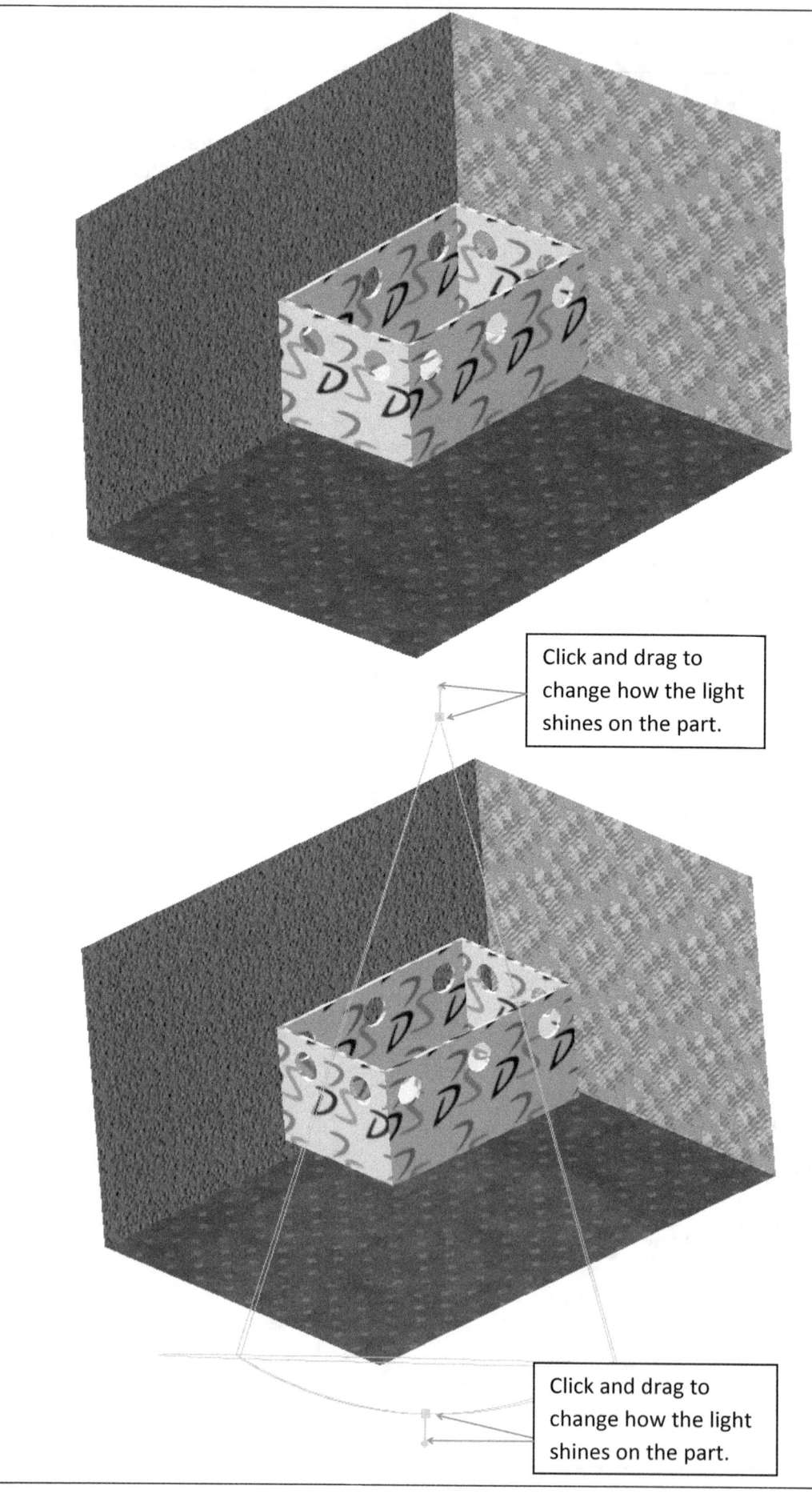

5. Click the Apply Material icon and apply patterns of your choice on the environment.
6. Click the Create Spot Light icon and add a light.
7. Click and drag the point and square to manipulate the light.

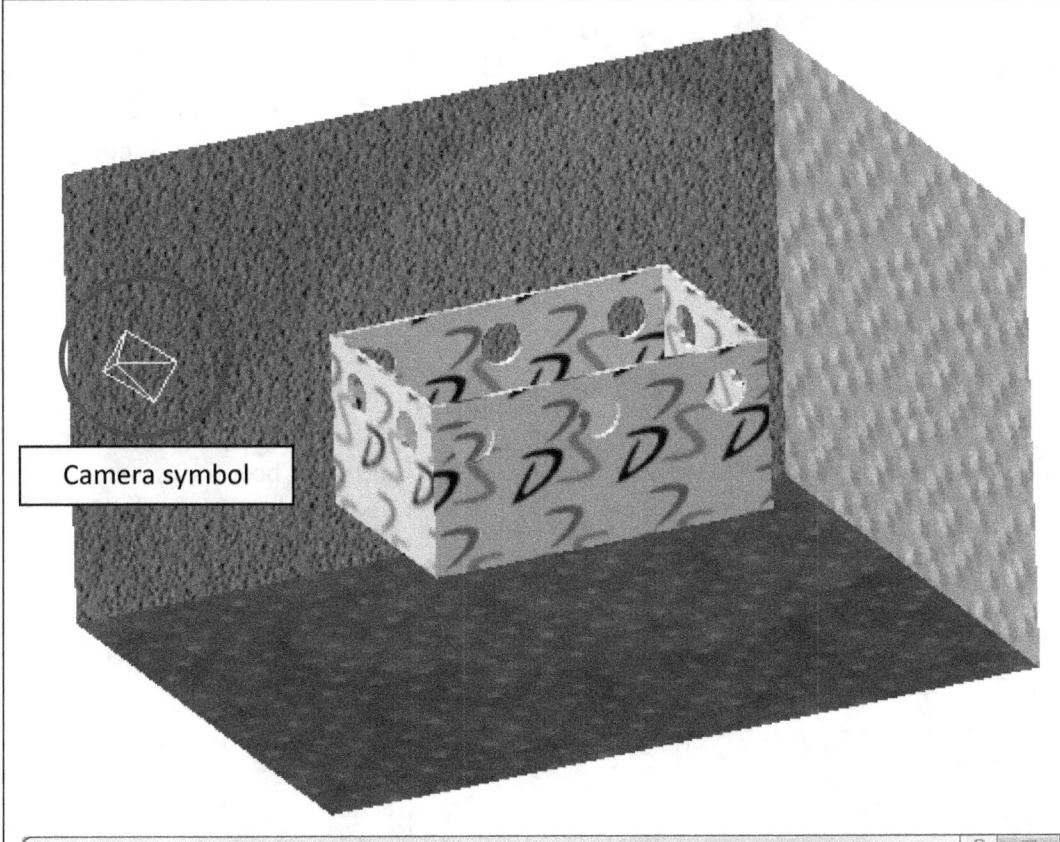

Camera symbol

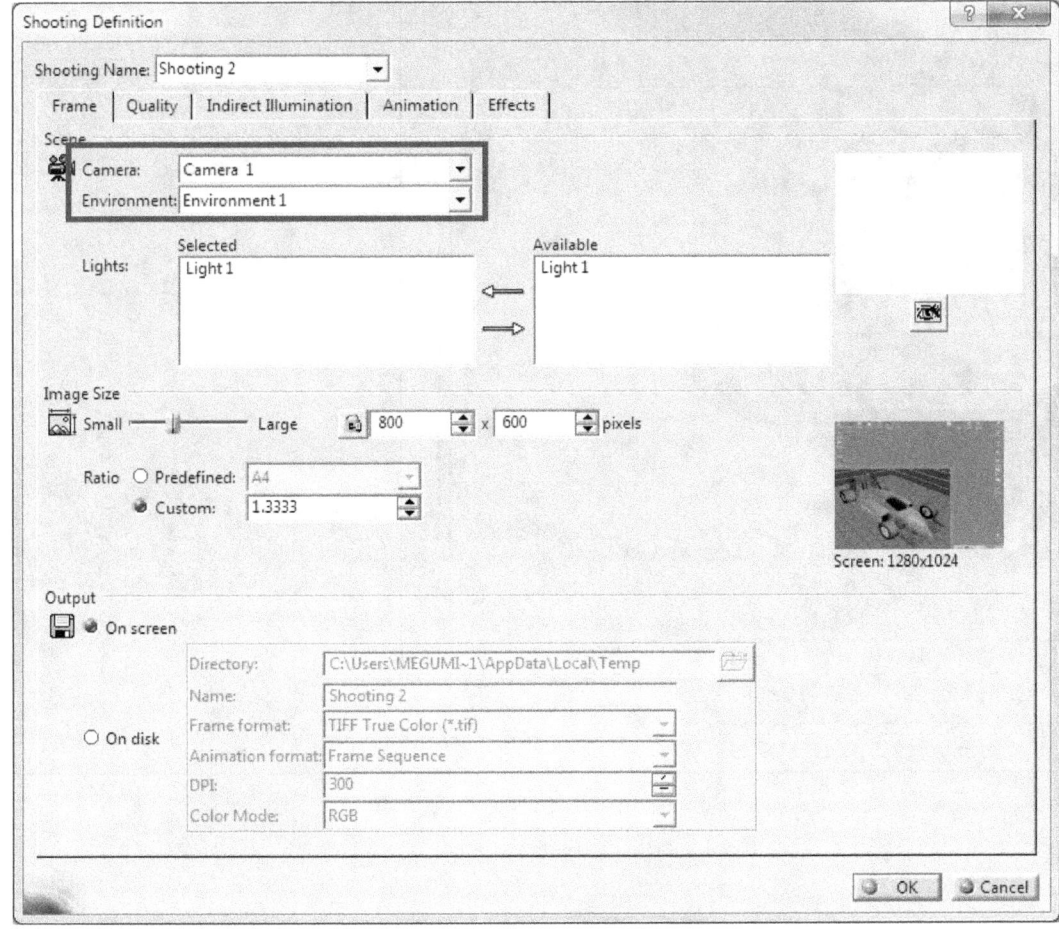

8. Click on the Create Camera icon 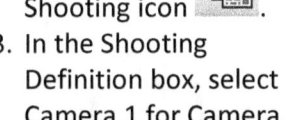 to add a camera angle.
9. Click on the Camera symbol to manipulate the camera angle and location.
10. Click the Quick Render icon to see how it looks.
11. Manipulate the light and camera to create a satisfying rendering.
12. Click on the Create Shooting icon.
13. In the Shooting Definition box, select Camera 1 for Camera, and Environment 1 for Environment.
14. Click OK.

93

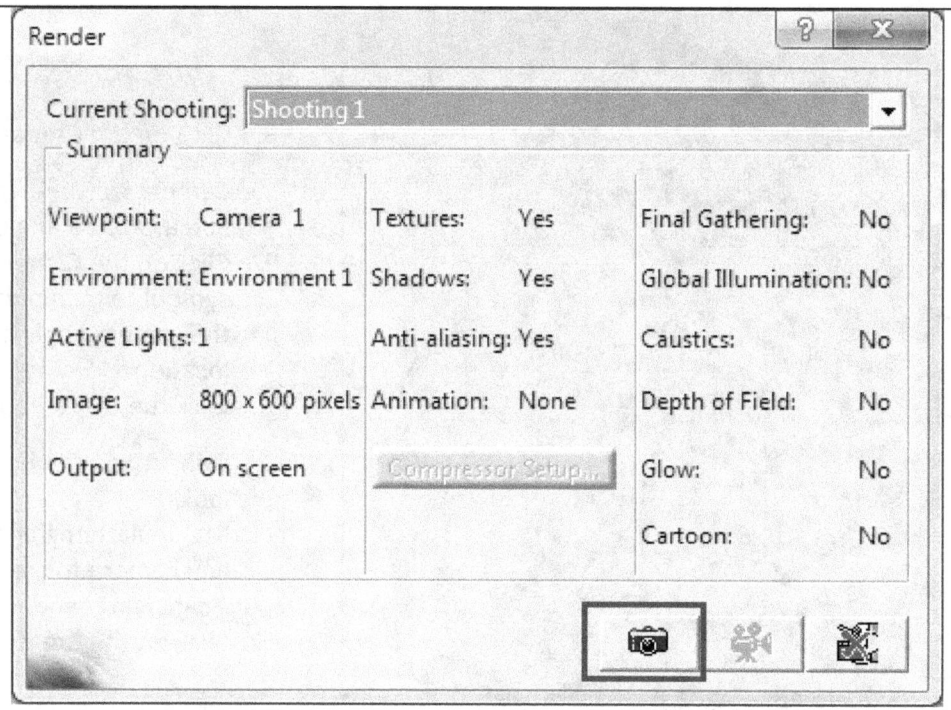

15. Click on the Rendering Shoot icon.
16. Make sure Shooting 1 is selected for the Current Shooting.
17. Click the camera icon in the box – this will generate the rendering.

The rendering will appear in a different box. If this is acceptable, click on the Save icon in the box to save as a high quality JPEG.

18. Save as "Render."

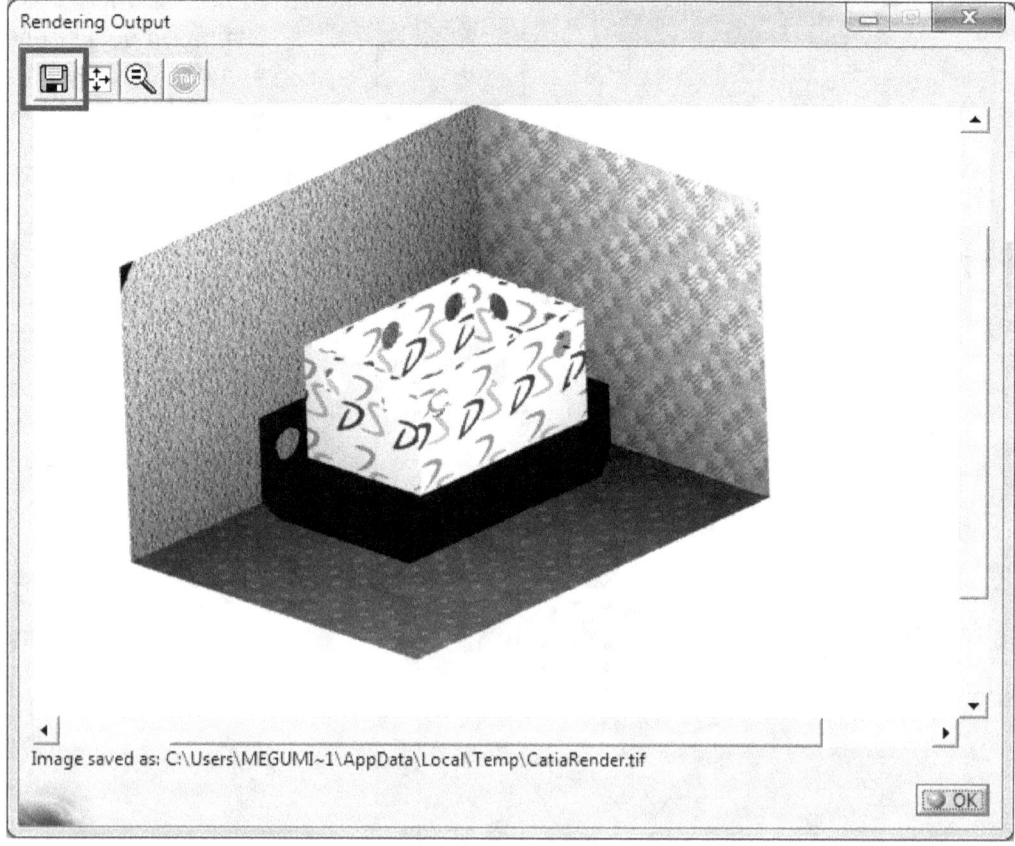

94

Another rendering method:

1. Open any part of your choice.
2. Click on the Photo Studio Easy tool icon . This changes the background.
3. Click on the Select Scene icon on the Render toolbar on the right bottom corner.
4. A scene box appears that has different scenes.
5. Select a scene.
6. Move/rotate/pan the part so that it sits in the scene appropriately.
7. Click on the Render icon.
8. If the rendering is acceptable, click on the save icon and save as a JPEG.

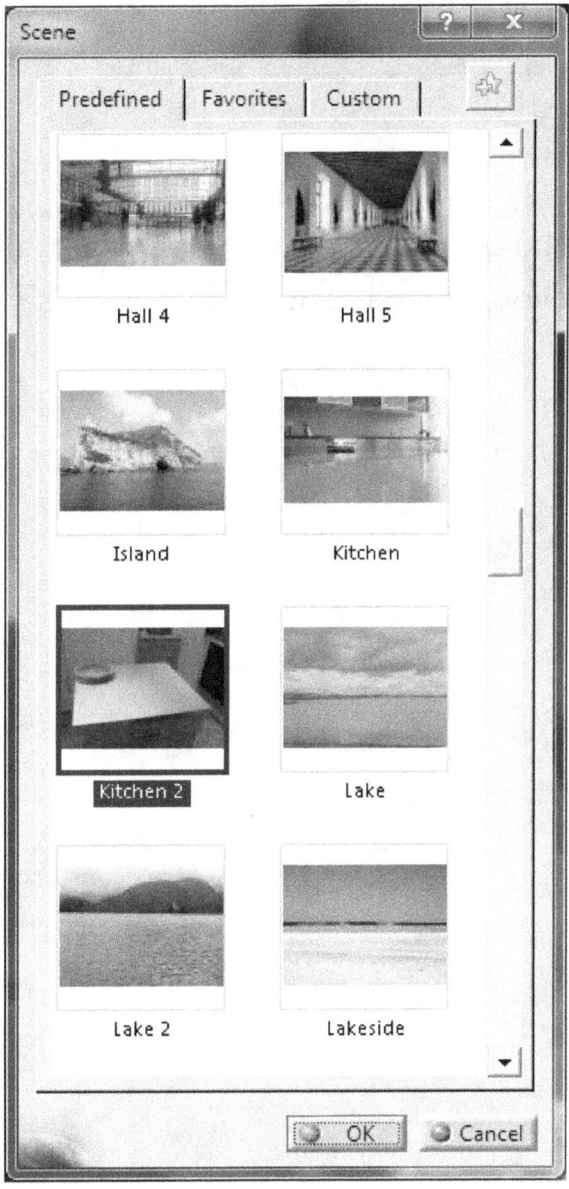

You can use the Define Rendered Area icon to select a certain area to render.

Chapter 7 Assignment

1. Create a rendering of a part of your choice with an environment, light and camera.
2. Create a rendering of a part of your choice with a scene.

Chapter 8 – Parametric Designing

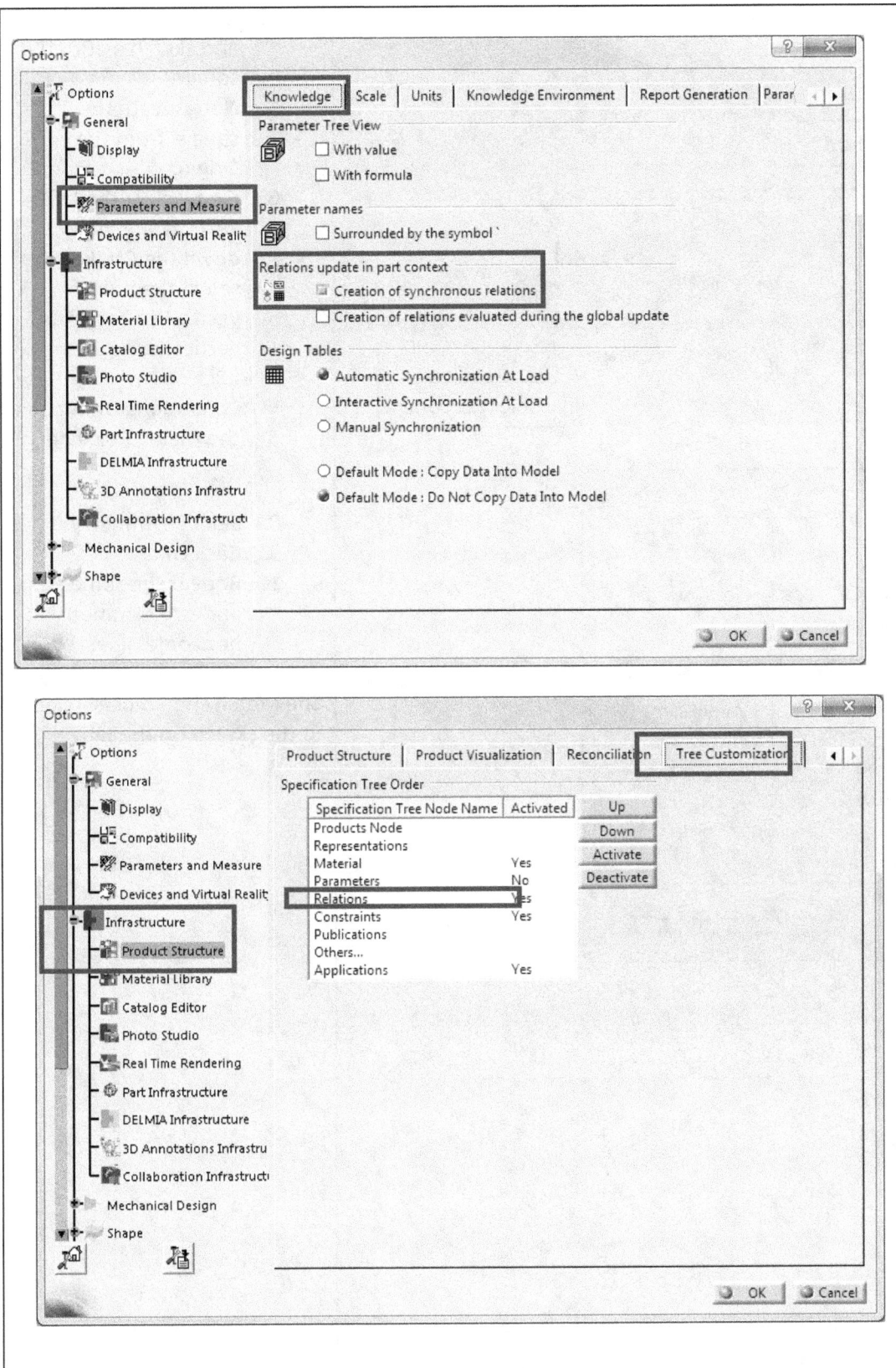

1. Open a new Part Design
2. Set it up in mm.
3. Make sure Relations are activated as shown on the left.

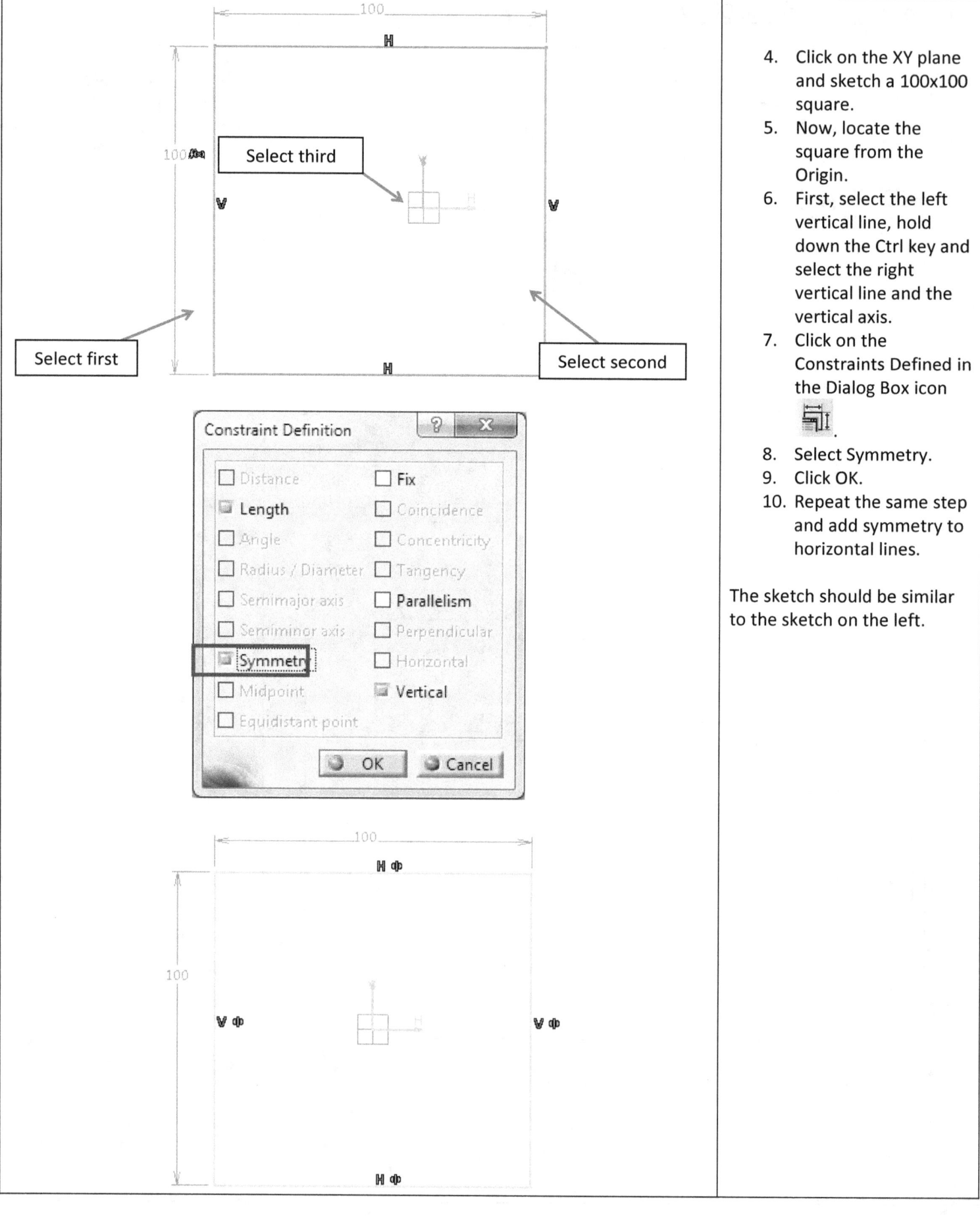

4. Click on the XY plane and sketch a 100x100 square.
5. Now, locate the square from the Origin.
6. First, select the left vertical line, hold down the Ctrl key and select the right vertical line and the vertical axis.
7. Click on the Constraints Defined in the Dialog Box icon.
8. Select Symmetry.
9. Click OK.
10. Repeat the same step and add symmetry to horizontal lines.

The sketch should be similar to the sketch on the left.

98

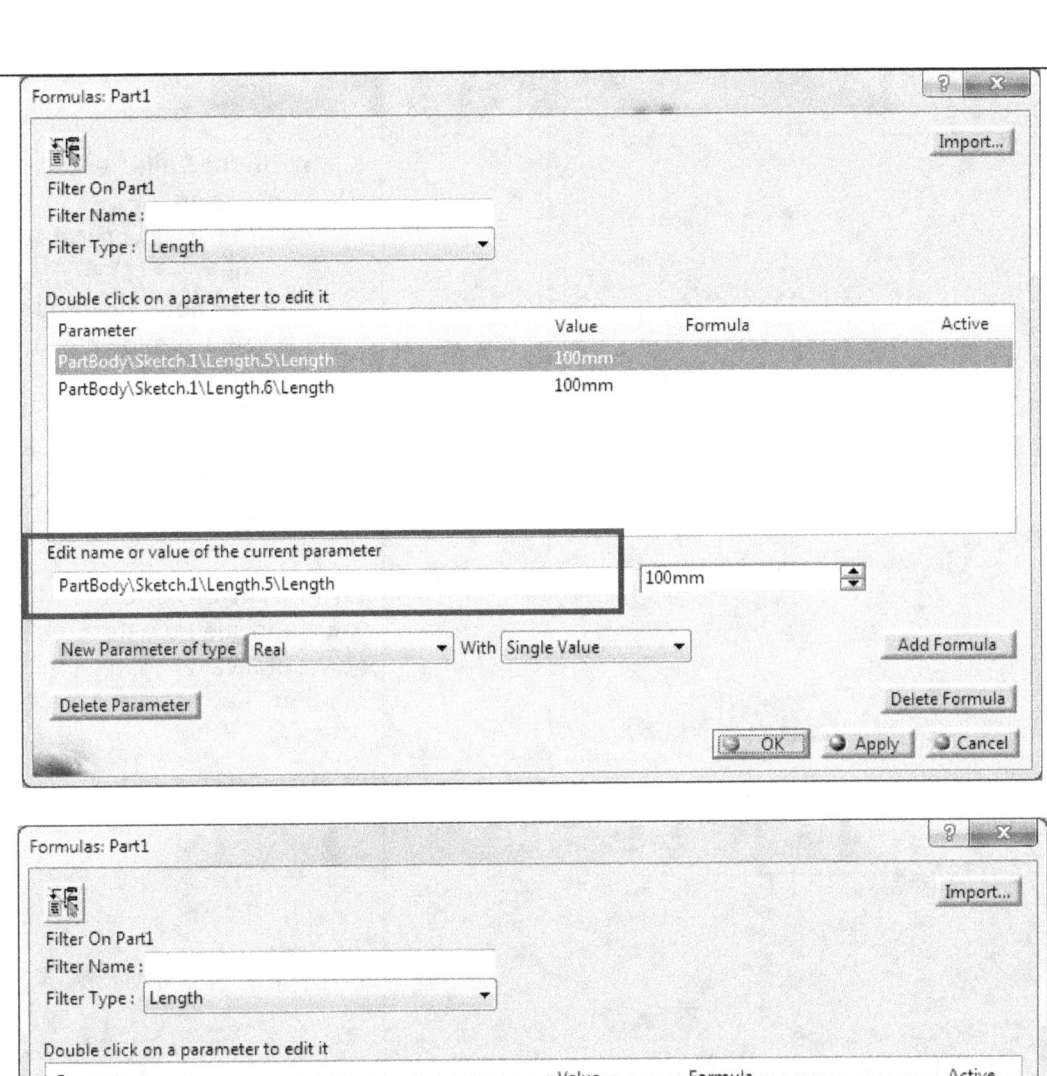

1. Click on the Formula icon $f(x)$.
2. Under Edit name or value of the current parameter, change the name to Horizontal line and Vertical line. In this case, Length 5 is Horizontal line and Length 6 is Vertical line.
3. It should be as shown on the left.
4. Click OK.

5. Double click on the horizontal 100 value.

99

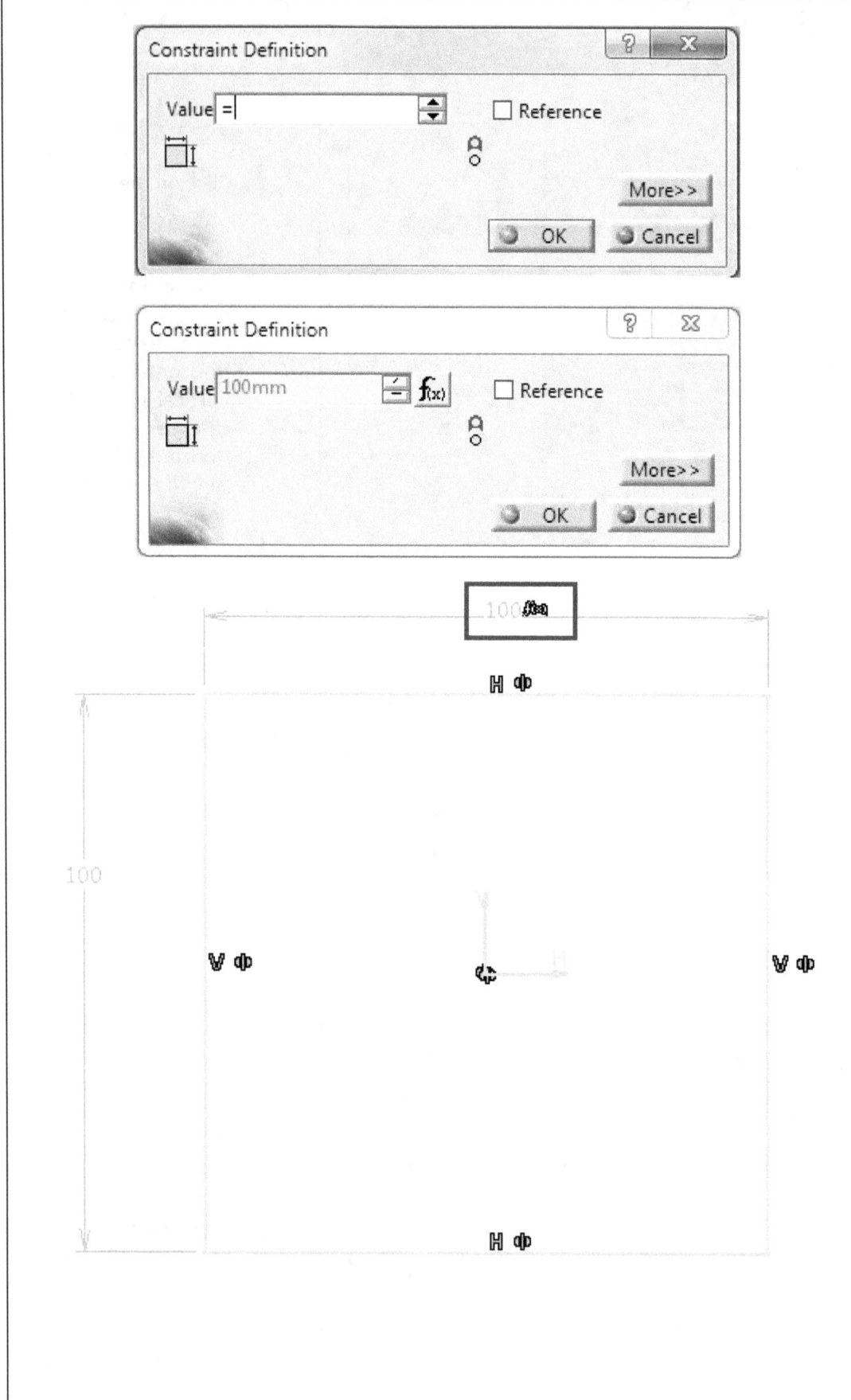

6. In the Value, enter "=" sign.
7. Then click vertical 100 value.
8. F(x) shows next to the value as shown on the left.
9. Click OK.
10. Horizontal 100 value has F(x) as shown on the left.

Try to change the vertical value and see how the horizontal value changes in relation.

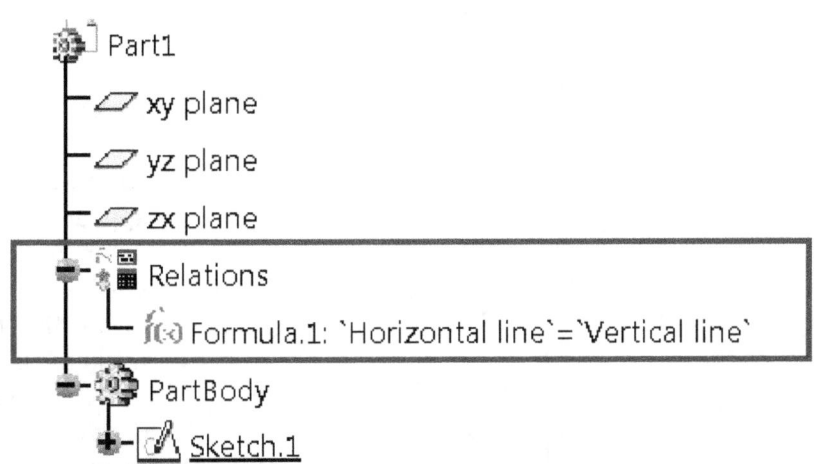

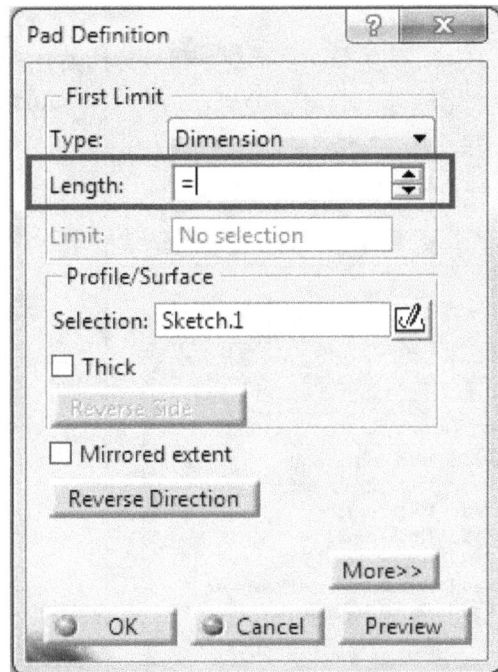

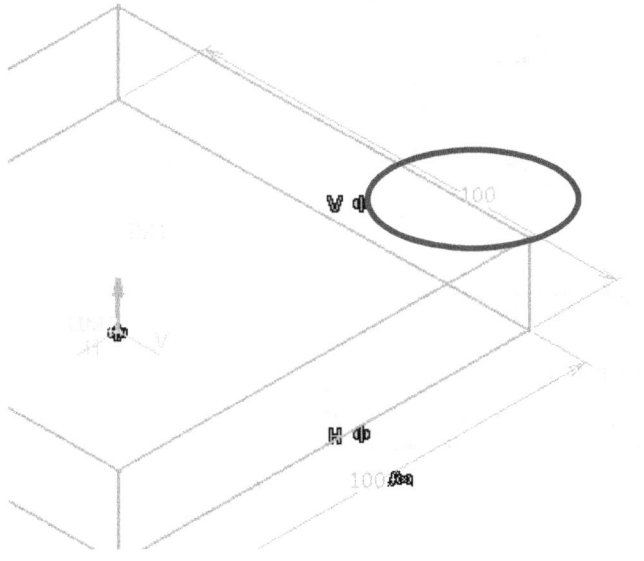

Exit the sketcher, and look at the Tree. Under the Relations, there should be a formula saying "Horizontal line = Vertical line." This means that the horizontal and vertical lines are always the same length.

Now let's add thickness with another formula.

1. Click Pad icon.
2. Click on the Sketch 1.
3. On the Pad Definition window, enter "=" sign in the Length box as shown on the left.
4. Then click on the Vertical line 100 in the Sketch 1.

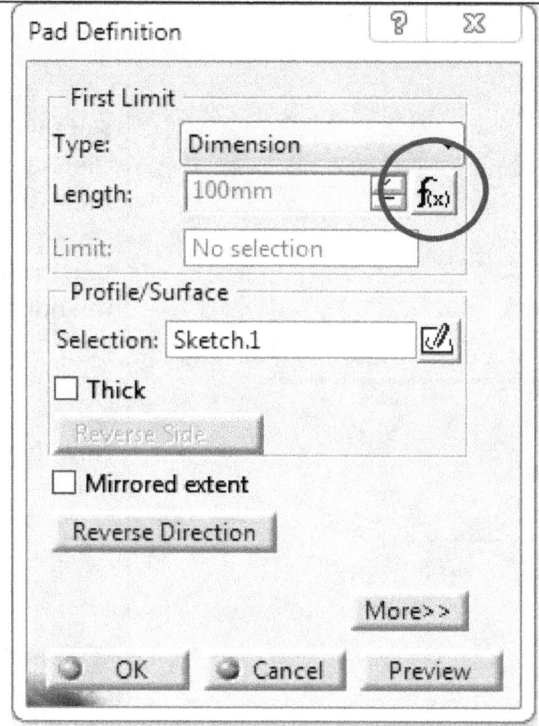

5. Click on the F(x) next to the Length.
6. Formula Editor box appears.
7. Enter /5 next to the Vertical line, which means that the thickness is always the 1/5 of length.
8. Click OK.
9. Again, click OK.

The Tree should be as shown on the left. Another formula should have been added.

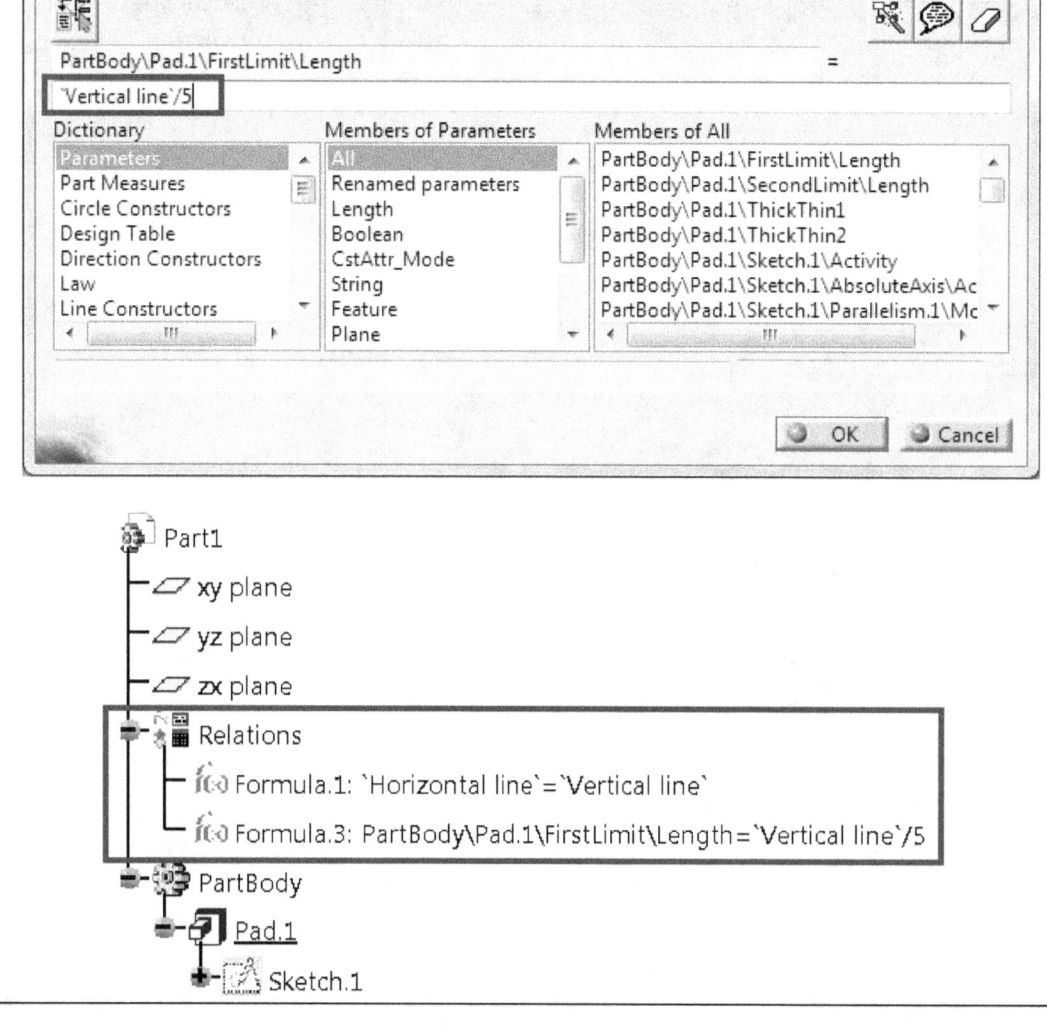

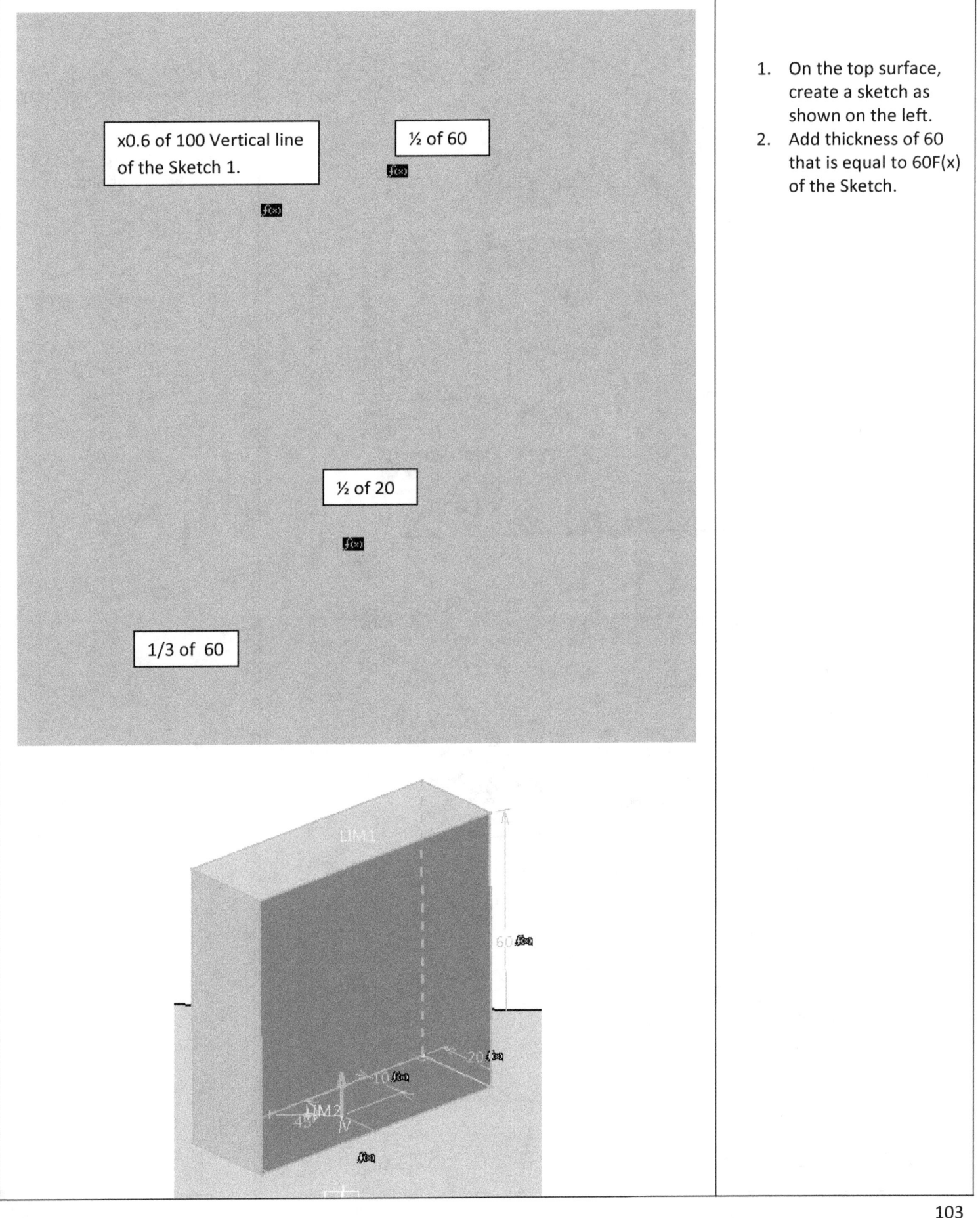

1. On the top surface, create a sketch as shown on the left.
2. Add thickness of 60 that is equal to 60F(x) of the Sketch.

3. Let's add a hole by clicking the Hole icon.
4. Click on the Face of Body 2 as shown on the left.
5. Select "Up to Next" when the hole definition pops up.
6. For Diameter, enter "=" sign first, then click on the Sketch 2 on the tree. Then select 60F(x).

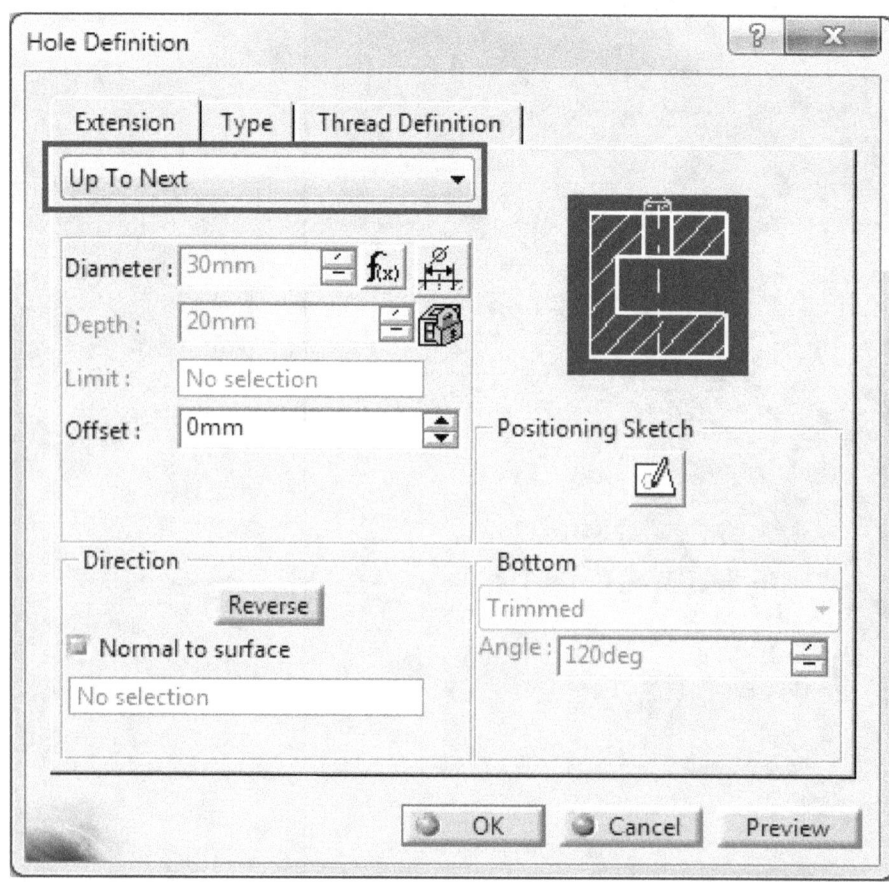

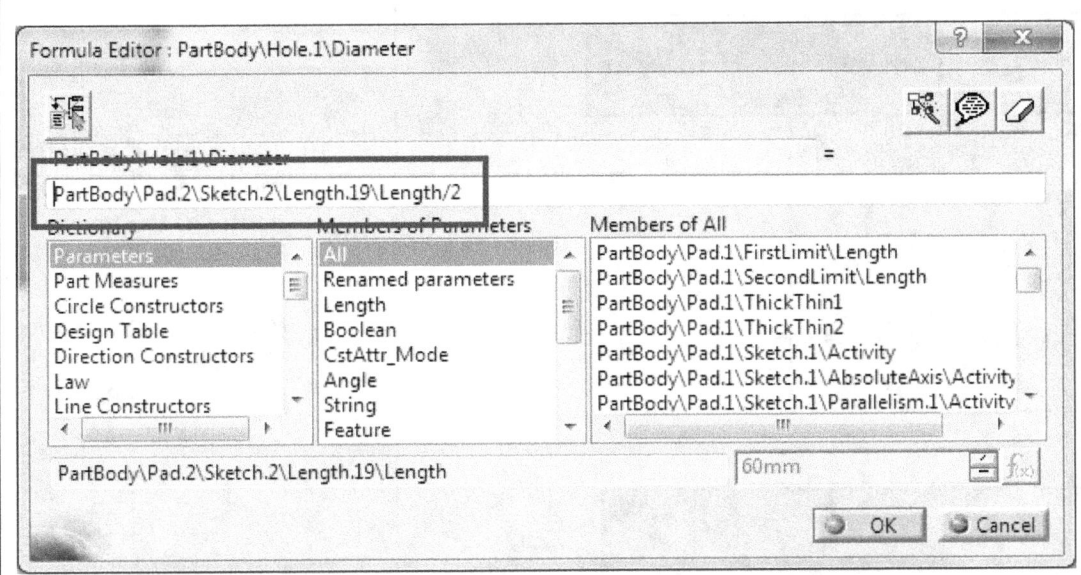

7. That appears as F(x) next to Diameter. Click the F(x) and enter "/2" as shown on the left.
8. Click OK.
9. Click on the Positioning Sketch icon on the Hole Definition box.

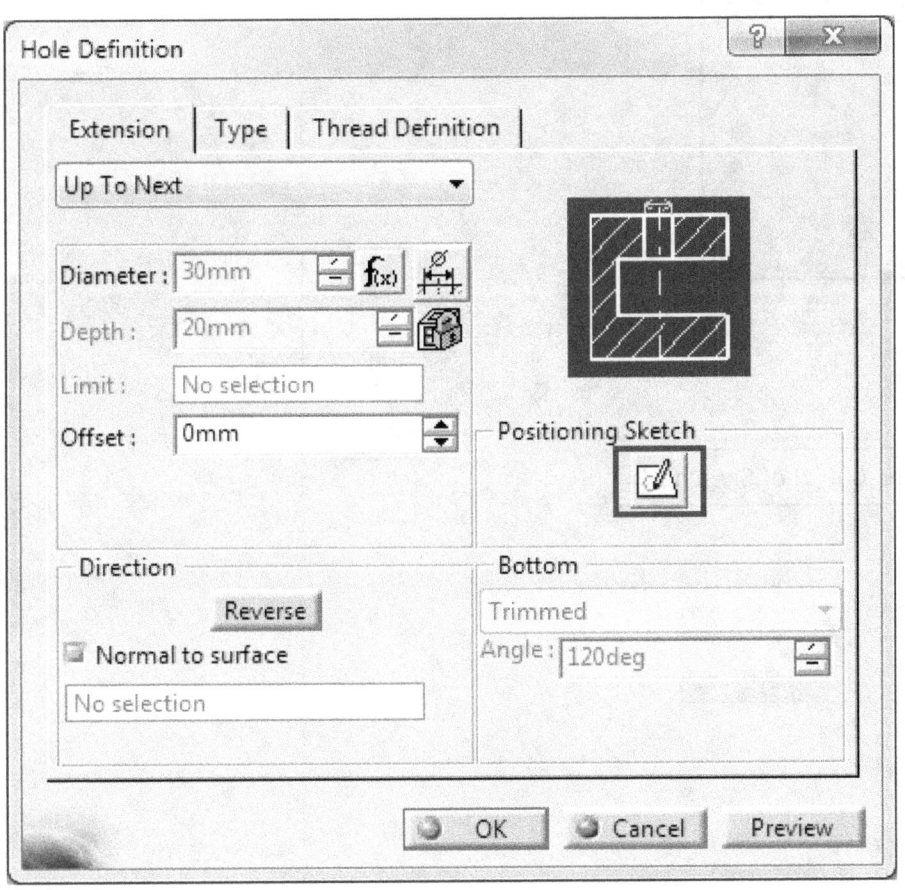

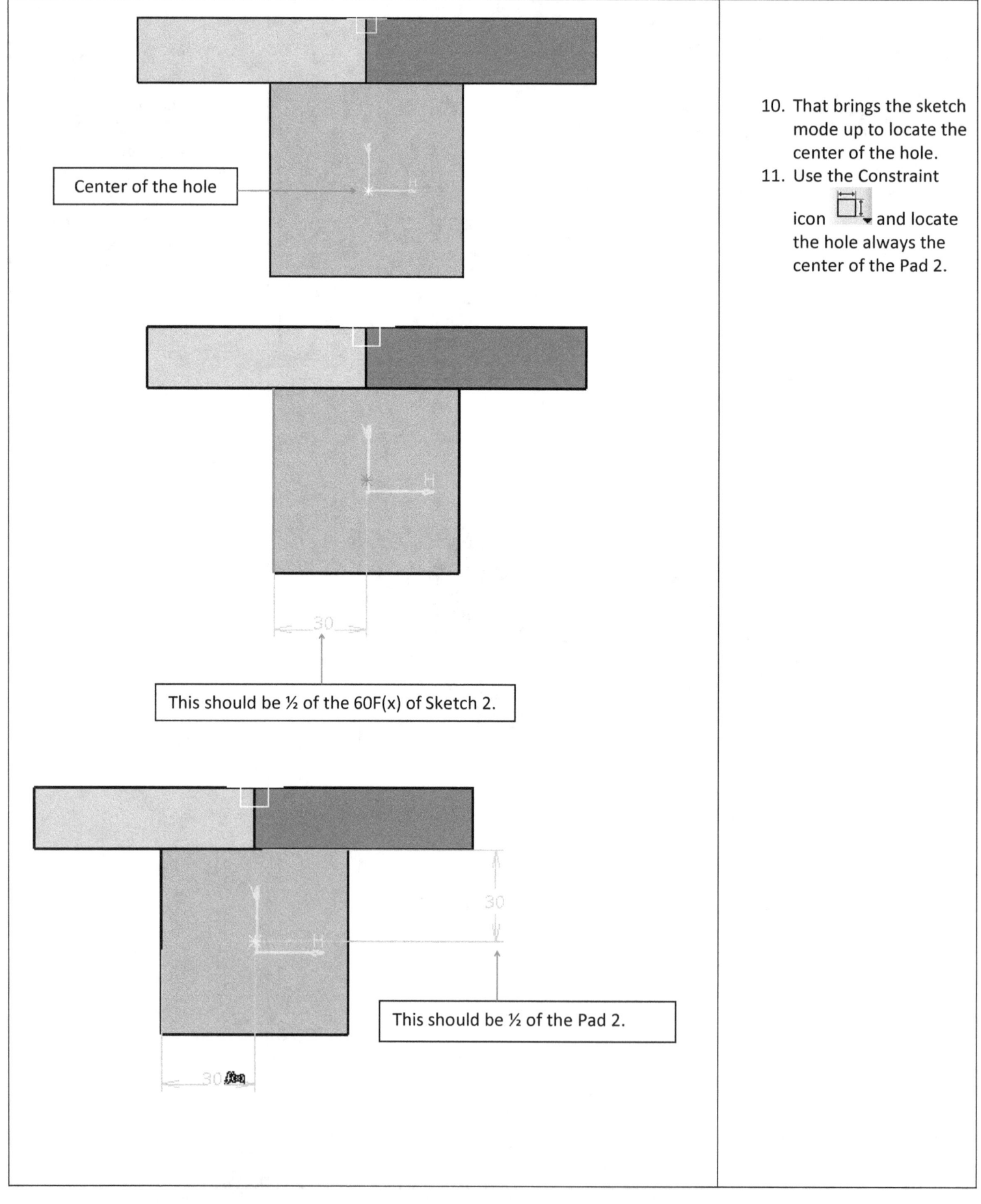

10. That brings the sketch mode up to locate the center of the hole.
11. Use the Constraint icon and locate the hole always the center of the Pad 2.

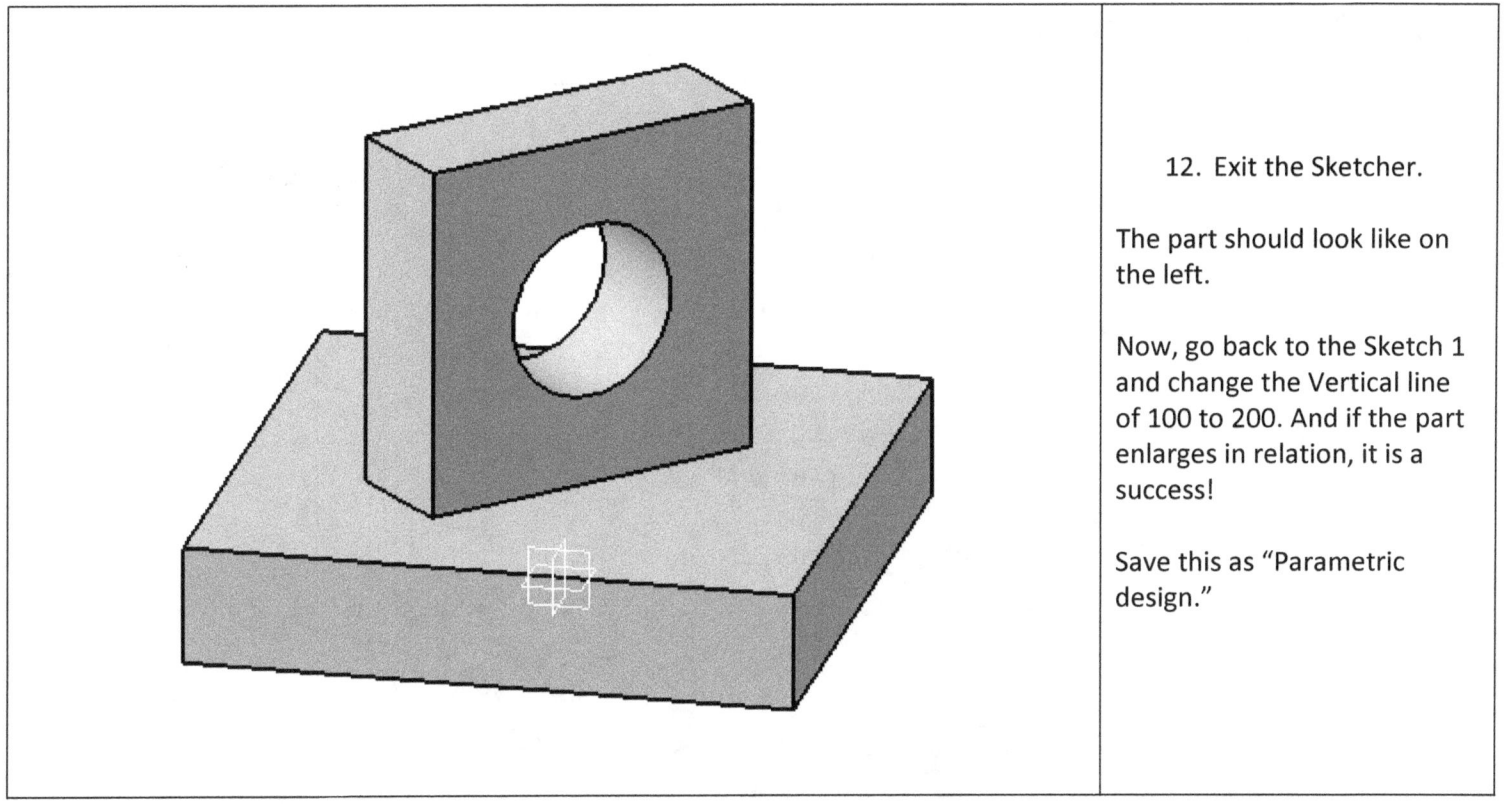

12. Exit the Sketcher.

The part should look like on the left.

Now, go back to the Sketch 1 and change the Vertical line of 100 to 200. And if the part enlarges in relation, it is a success!

Save this as "Parametric design."

Chapter 8 Assignment.

1. Open Chapter 2 Exercise 2 and add formulas so that it enlarges/shrinks relatively.

Chapter 9: Advanced Techniques

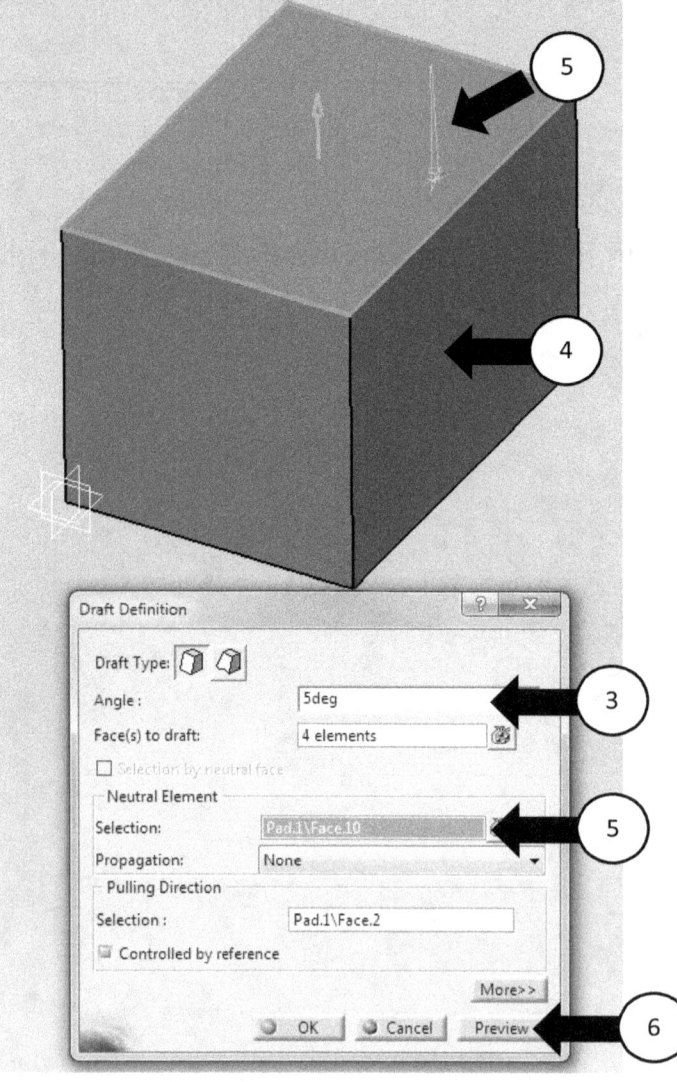

1. Create a box similar to the left.
2. Click on the Draft Angle icon
3. Set the angle.
4. Select 4 side surfaces.
5. Click the top surface.
6. Click Preview.

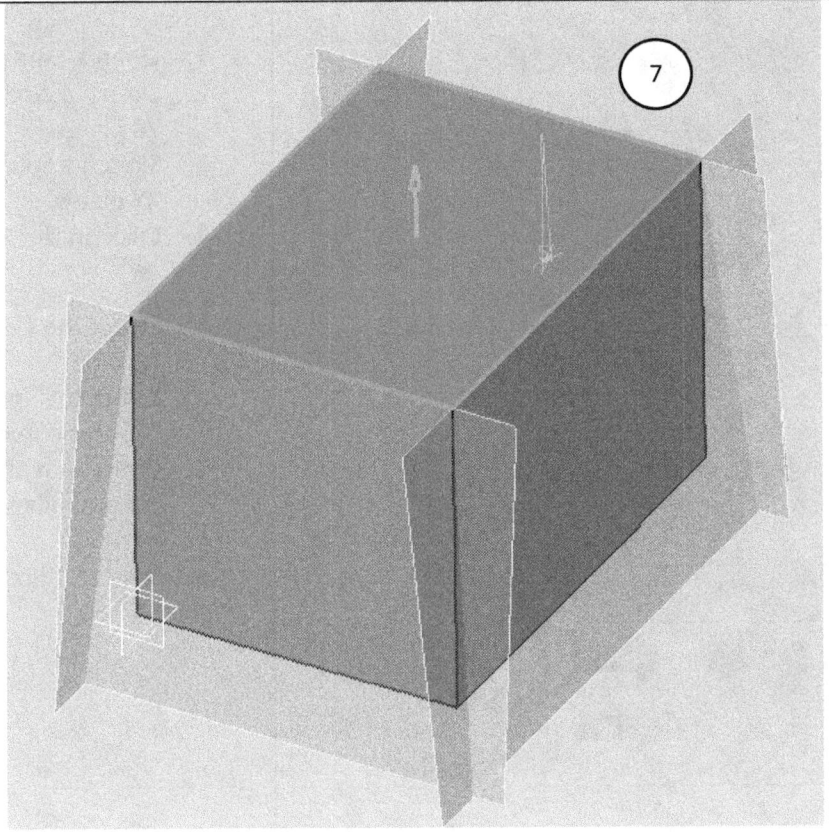

7. It should look similar to the left.
8. Click OK.
9. Save as "Draft Angle"

The box now has Draft Angle to the side surfaces.

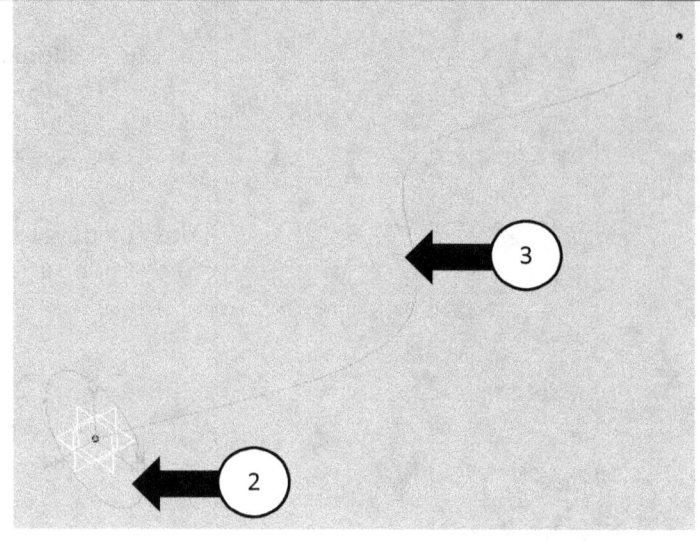
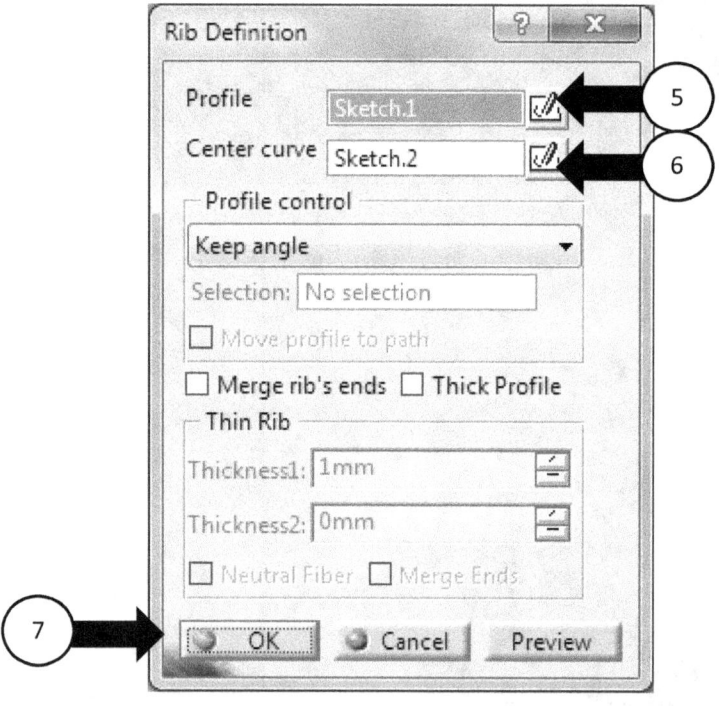

1. Open a new part.
2. Sketch a circle on the YZ plane.
3. Sketch a spline on the XY plane.
4. Click on the Rib icon.
5. Select the circle for "Profile."
6. Select the spline for "Center curve."
7. Click OK. It should looks similar to the left.
8. Save as "Rib."

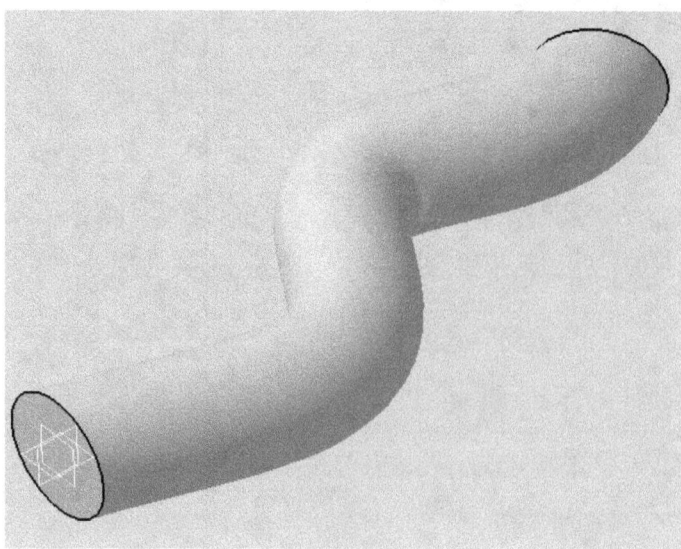

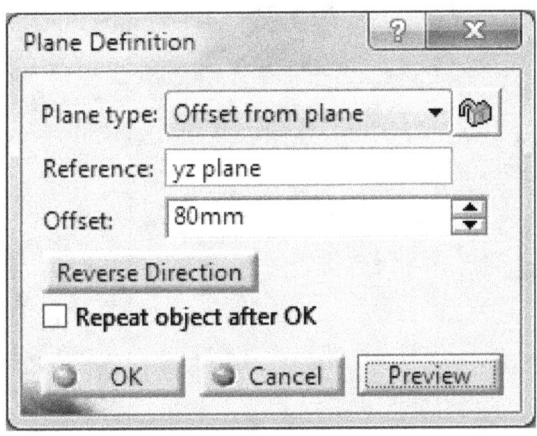
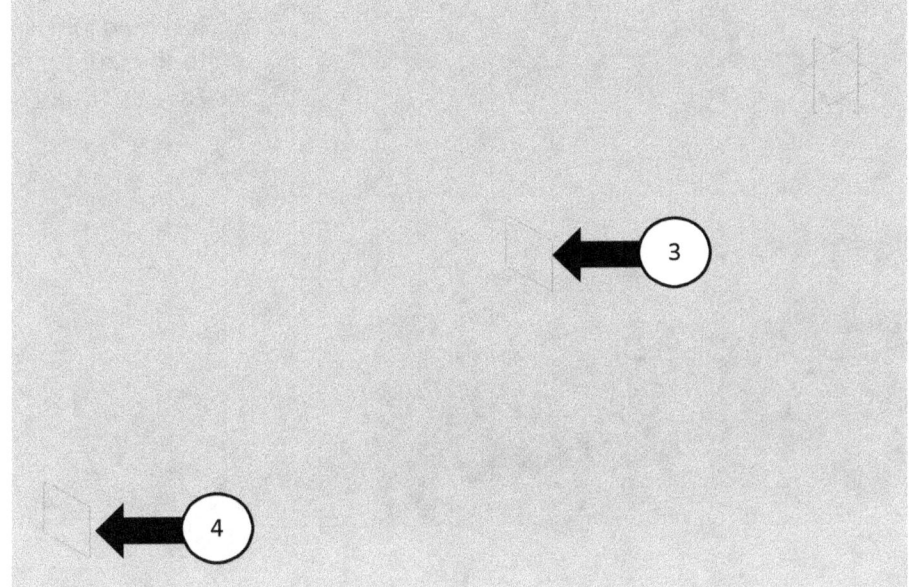
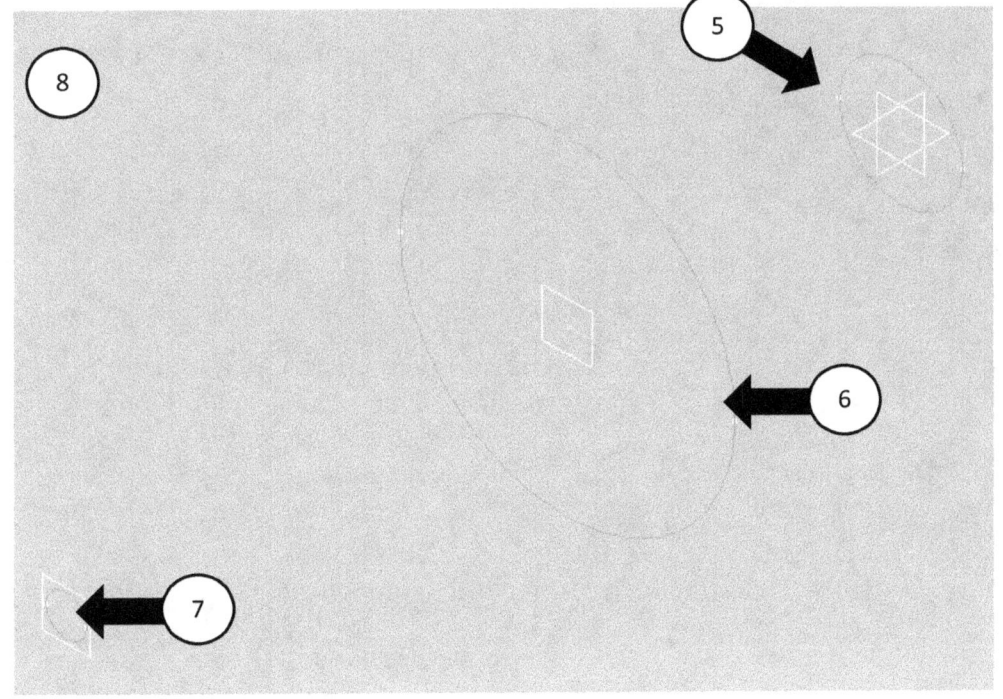

1. Open a new part.
2. Click on the Plane icon.
3. Offset 80 mm from the YZ plane. (Plane 1)
4. Offset 200 mm from the YZ plane. (Plane 2)
5. Click on the YZ plane and create a circle, Ø30 in the very center.
6. Create a circle, Ø80 in the very center on the Plane 1.
7. Create a circle, Ø10 in the very center on the Plane 2.
8. It should look similar to the left.

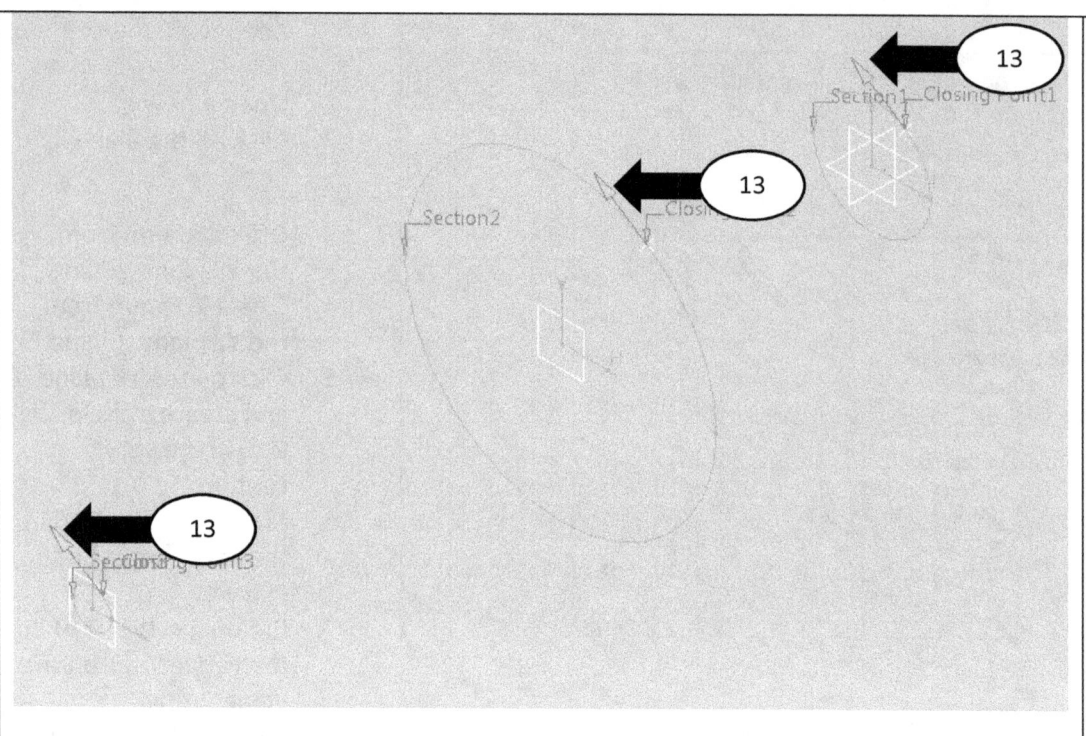

9. Click on the Multi-Sections Solid icon .
10. Select the Ø30 circle.
11. Select the Ø80 circle.
12. Select the Ø10 circle.
13. Ensure all the arrows are facing the same direction.
14. Click OK.
15. It should look similar to the left.
16. Save as "Multi-solid."

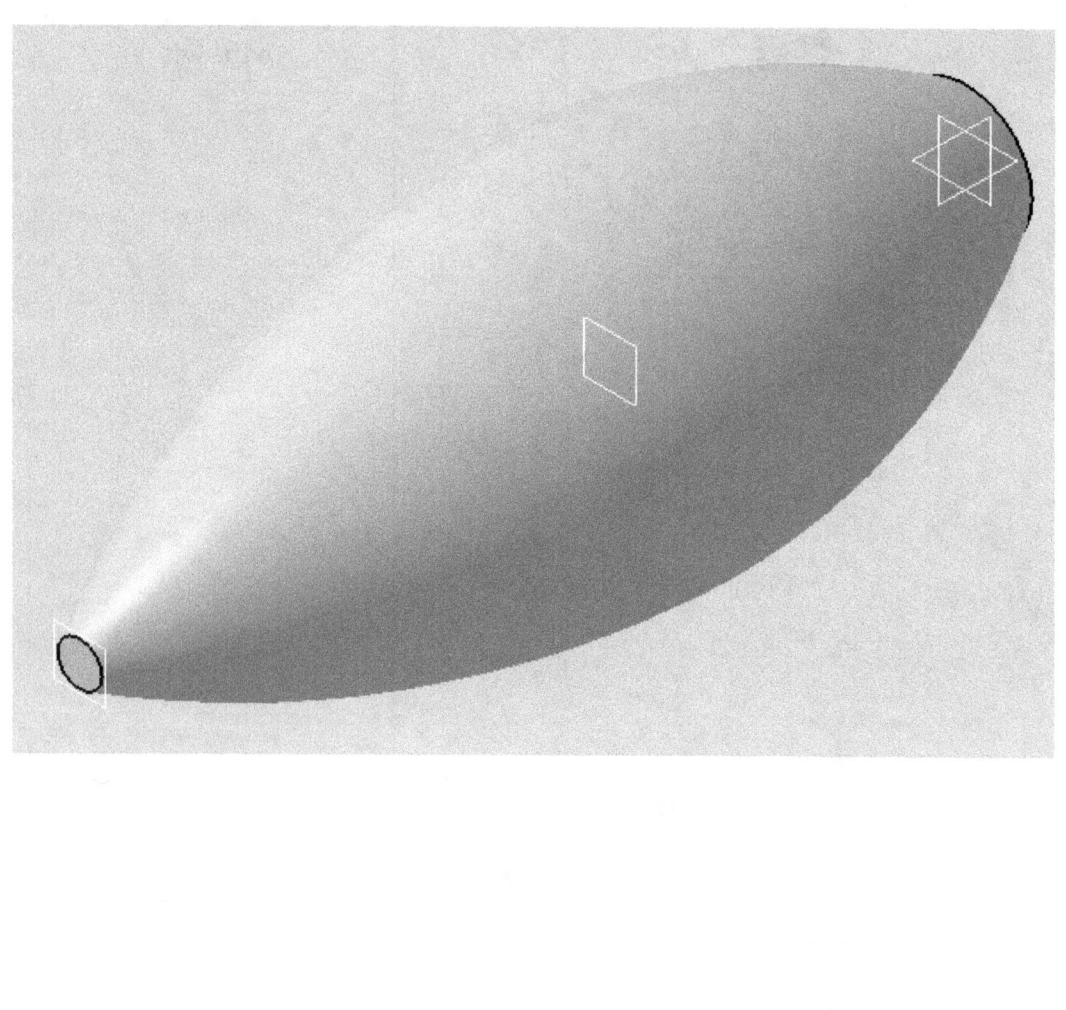

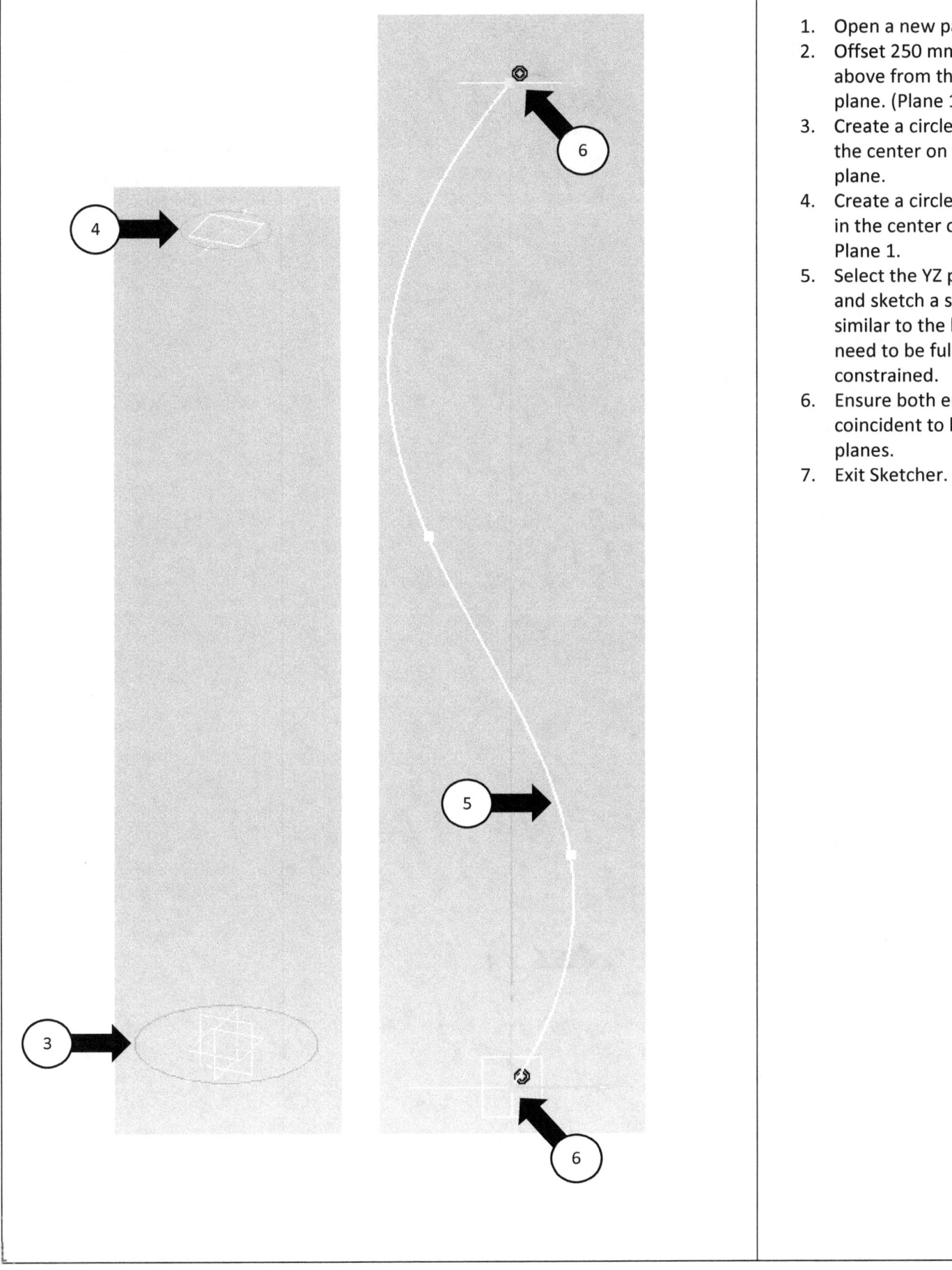

1. Open a new part.
2. Offset 250 mm to above from the XY plane. (Plane 1)
3. Create a circle, Ø40 in the center on the XY plane.
4. Create a circle of Ø20 in the center on the Plane 1.
5. Select the YZ plane and sketch a spline similar to the left. No need to be fully constrained.
6. Ensure both ends are coincident to both planes.
7. Exit Sketcher.

8. Click on the Multi-Sections Solid icon .

9. Select both circles. Ensure the arrow direction is the same.

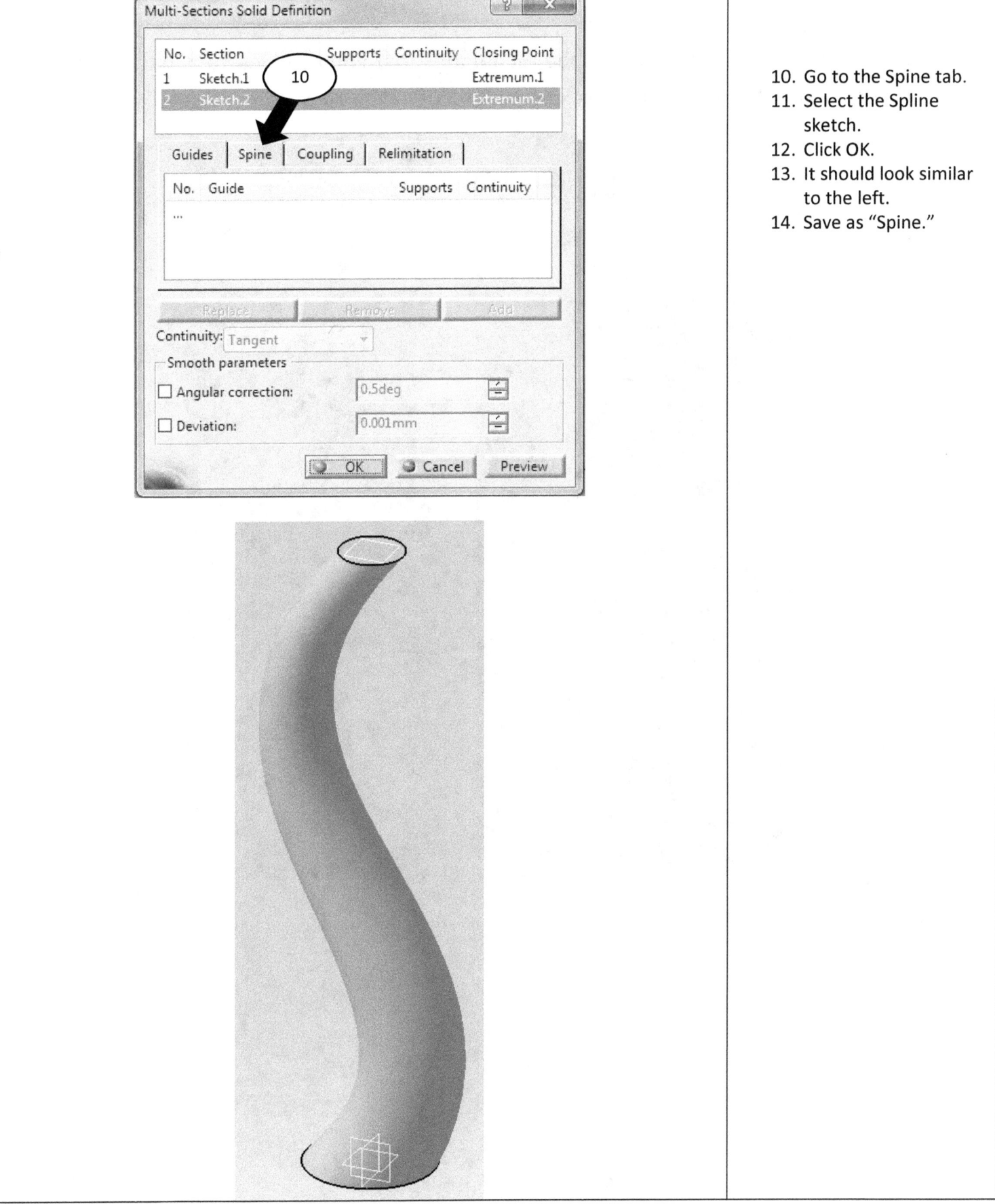

10. Go to the Spine tab.
11. Select the Spline sketch.
12. Click OK.
13. It should look similar to the left.
14. Save as "Spine."

Chapter 9 Assignment

1. Create the following toy using all the techniques that you have learned in this book. Do not worry about screws in the back side, or colors/materials.

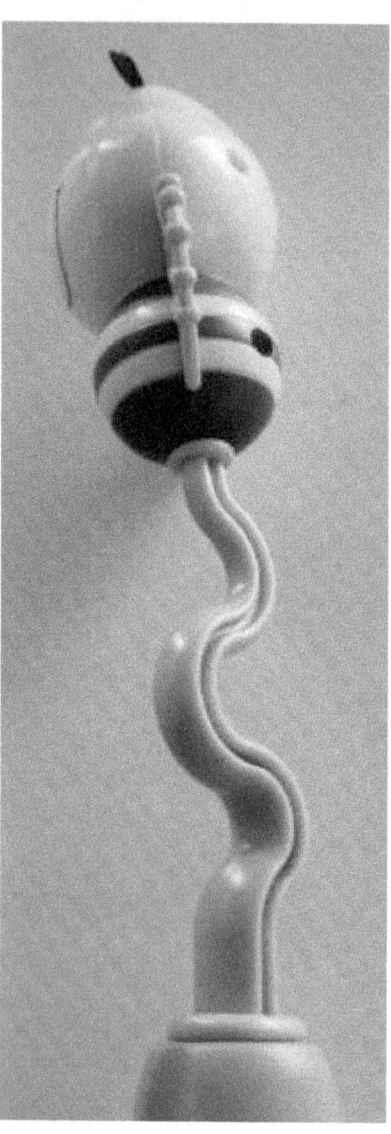

© Megumi Leatherbury

www.ingramcontent.com/pod-product-compliance
Lightning Source LLC
Chambersburg PA
CBHW080930170526
45158CB00008B/2235